ry Grain
ɔn Systems
ᴠesign Handbook

This handbook was prepared under the direction of the Dry Grain Aeration Systems Design Handbook Committee of MidWest Plan Service.

Dry Grain Aeration Systems Design Committee
Ken Hellevang, North Dakota State University (chair)
Leslie Backer, North Dakota State University
Roger Brook, Michigan State University
Joseph Harner, Kansas State University
David Jones, University of Nebraska
Dirk Maier, Purdue University
Bill Peterson, University of Illinois (retired)
William Wilcke, University of Minnesota

Ken Hellevang was author or coauthor of the sections on fans and hopper bins and of portions of the bin and flat storage design sections. He provided a major rewrite of the first draft of the book and did the final editing for this revised edition. Roger Brook wrote portions of the flat storage section, assisted with the final review of the first edition, and helped edit this revision. Bill Peterson wrote the section on controls. William Wilcke helped edit both the first edition and this revised edition.

Others who reviewed and participated in developing a design consensus were Dirk Maier, Joseph Harner, Leslie Backer, and David Jones. Larry Van Fossen, formerly with Iowa State University and now deceased, and Mike Veenhuizen, formerly with MidWest Plan Service, wrote the original manuscript for the first edition. Richard Spray, formerly of Clemson University and now deceased, reviewed the first edition.

MidWest Plan Service
122 Davidson Hall
Iowa State University
Ames, IA 50011-3080

MidWest Plan Service
A Foundation of Knowledge

Table of Contents

List of Figures

List of Tables

List of Examples

Table of Contents

List of Figures

List of Tables

List of Examples

Publication Overview

This design handbook is intended for a technically oriented audience interested in learning about the art and science of aeration system design. It provides guidelines for selecting, sizing, locating, and evaluating grain aeration systems, and it presents design examples of commonly used systems. It **does not** include design information for moving air through wet grain to hold it safely until it is dried, or for cooling hot grain coming from a dryer.

Design values in this handbook are conservative minimums based on the experience and training of the authors. Designers choosing to use values other than those expressed in the book do so based on their experiences and training and should recognize that their results may not be consistent with the results this publication describes.

Many factors affect design values:
- Type of grain.
- Depth of grain.
- Time available to cool/warm grain.
- Cost of aeration system.
- Energy costs.
- Practical duct size.
- Management goals.

Proper design of an aeration system is important, but correct operating techniques are equally important. Often times, management skill can compensate for limitations of a system, but seldom does system design compensate for faulty management. Contact the local Extension Service for management information.

Basic Aeration Considerations

Aeration is a management process that forces air through dry grain to control grain temperatures in storages. Aeration helps maintain grain quality by limiting moisture movement in the grain caused by temperature differentials. It also limits the potential for mold production and insect activity.

The purpose of aeration is to control grain temperature and maintain moisture content for long-term storage (Table A.1). Spoilage will likely occur when aeration airflow rates are used to dry grain. However, the higher airflow rates used for drying grain can be used for aeration.

The major issues to consider in aeration system design are the following:

- System components.
- Airflow rate.
- Aeration cycle time.
- Airflow distribution.
- Airflow direction.

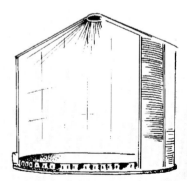

System Components

A dry grain aeration system, Figure 1.1, has these components:

- A fan and motor to move the necessary airflow at the required static pressure.
- Transition and/or supply ducts to connect the fan to the air distribution system.
- An air distribution system; including perforated ducts, floor sections, or floors under the grain.
- Exhaust vents for positive pressure systems or intake vents for negative pressure systems.
- Controls to operate and regulate fan operation.

For effective performance of the system, prevent moisture from entering the structure. Slope the ground away from the storage, and seal structural joints at the foundation or footing. Slope the foundation or footings so water does not pond at the joint.

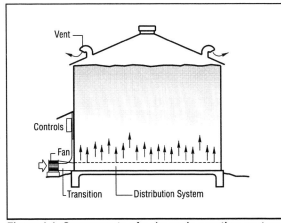

Figure 1.1. Components of a dry grain aeration system.

Airflow Rate

The selected airflow rate for a dry grain aeration system depends on the
- Type of grain stored.
- Air distribution system preferred.
- Management procedures used.
- Climatic conditions.
- Desired time to warm or cool the grain.

An airflow rate of about 0.10 cfm/bu is widely used in the Midwest to design aeration systems for farm-stored shelled corn, soybeans, and grain sorghum. A higher airflow rate is often selected for wheat and other farm-stored grains that are harvested during the summer and need to be cooled when climatic conditions limit available cooling time. Lower airflow rates may be used

in flat storages to limit the duct size. Airflow rates of 0.03 to 0.05 cfm/bu are often selected for grain depths exceeding about 50 feet because higher airflow rates result in excessive static pressure and horsepower requirements.

Total required airflow, cfm, for the stored grain based on the airflow rate per bushel is calculated by Equation 1.1.

Equation 1.1 Total Required Airflow

$Q = (GV)(AR)$

Where: Q = Total required airflow, cfm
GV = Volume of grain, bu
AR = Airflow rate, cfm/bu

Example 1.1 Estimated total required airflow.
Estimate the total required airflow for 5,000 bushels of shelled corn if the desired airflow rate is 0.10 cfm/bu.
    Using Equation 1:
Q = 5,000 bu x 0.10 cfm/bu
    = 500 cfm

Aeration Cycle Time

During grain cooling or warming, temperature change occurs in a zone, or front, several feet thick that moves through the bin in the direction of airflow. An aeration cycle is the time it takes for one cooling or warming zone to pass completely through the grain, Figure 1.2.

The desired time to warm or cool the grain is an important factor. A shorter time permits completing aeration during periods of limited appropriate air conditions, but initial system cost is higher. Fan operating cost and the availability of favorable off-peak rates are additional factors to consider. The length of an aeration cycle depends on

- Amount and distribution of foreign material.
- Airflow rate, cfm/bu.
- Test weight of the grain. (Heavier grain has more heat to be removed.)
- Airflow distribution system.
- Initial temperature and temperature distribution.
- Time of year.

The length of an aeration cycle is only an estimate because specific heat varies with the type of grain. Equations 1.2 and 1.3 can be used to estimate hours per cycle to aerate grain. For more information consult the sources given in Appendix B, **References**. See Foster and Stahl, 1959; Kline and Converse, 1961; Holman, 1960.

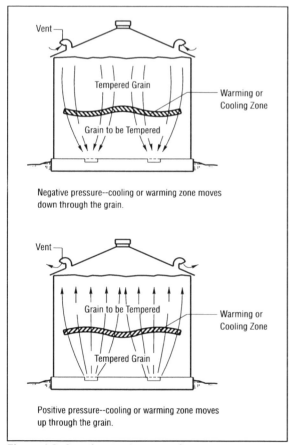

Figure 1.2. Aerating to change grain temperature.
Grain temperature changes over a period of hours as a
warming or cooling zone moves through the storage.

Negative pressure--cooling or warming zone moves
down through the grain.

Positive pressure--cooling or warming zone moves
up through the grain.

Equation 1.2 Aeration Time for 60 lb/bu Grain

$$AT = \frac{15}{AR}$$

Where: AT = Aeration time, hrs
AR = Airflow rate, cfm/bu

Example 1.2 Calculating aeration time for 60 lb/bu grain
How long does an aeration cycle take in 60 lb/bu grain with an airflow rate of 0.10 cfm/bu?

$$Hours/cycle = \frac{15}{0.10}$$

Hours/cycle =150 hrs. (or 6.25 days)

Equation 1.3 Aeration Time for All Grain

$$AT = \left(\frac{TW}{60}\right)\left(\frac{15}{AR}\right)$$

Where: AT = Aeration time, hr
TW = Test weight, lb/bu

Example 1.3 Calculating aeration time for all grain.
What is the estimated time for one aeration cycle through oats at 32 lb/bu using an airflow of 0.1 cfm/bu?

$$Hours/cycle = \left(\frac{32}{60}\right)\left(\frac{15}{0.10}\right)$$

$$= (0.53) \times (150\ hr)$$
$$= 80\ hr\ (or\ 3.3\ days)$$

Approximate aeration cycle times for other airflow rates and test weights are provided in Table 1.1. Several cycles are needed to control grain temperature during the year.

The rate of grain temperature change (hours per cycle) is directly proportional to the airflow rate, cfm/bu. Doubling the airflow rate cools or warms the grain about twice as fast, but doubling airflow rate requires about five times the fan motor horsepower.

Table 1.1. Approximate times for one aeration cycle, hours.

Times are based on 10 F to 15 F temperature changes in the Midwest. Times vary with air conditions, amount of foreign material, airflow rate variation within the grain, and warming vs. cooling.

Airflow rate, cfm/bu	Grain test weight, lb/bu			
	32	48	56	60
	Aeration Cycle, hr			
0.03	240	360	420	450
0.05	160	240	280	300
0.10	80	120	140	150
0.20	40	60	70	75
0.25	32	48	56	60
0.50	16	24	28	30
0.75	11	16	19	20
1.00	8	12	14	15

Airflow Distribution

Airflow distribution is not uniform throughout grain. Among the factors that affect airflow distribution are these:

- Pile geometry, especially grain depth.
- Equipment used in the aeration system; see Chapter 2, **System Components**.
- Variation in cleanliness and density (test weight) of grain.
- Distribution of broken grain and foreign material (fines).

Nonuniform airflow means there is an **actual** airflow rate (delivered airflow rate) that varies throughout the grain; and a design airflow rate. Effective system design and skilled management minimize the difference between design airflow and actual (delivered) airflow. The variation in actual airflow rate is what makes management a crucial ingredient to successful grain storage.

Nonuniform airflow causes

- Longer operation time. The fan must continue operating to cool or warm the portion of grain with the lowest airflow rate, while part of the airflow will continue to move through grain that has already been cooled or warmed.
- Difficulty in determining when an aeration cycle is complete.
- Increased operation expenses.

Airflow Direction

Aeration distribution systems are classified as either positive pressure systems (push systems) or negative pressure systems (suction systems). A positive pressure system pushes air up through the grain. A negative pressure system pulls air down through the grain.

Although one system may be more appropriate in a particular application, either can be used effectively for aeration when equipment is properly installed and managed. Fans, when installed correctly, perform the same whether pushing or pulling air.

In a **positive pressure system**

- Air distribution in flat storage duct systems is more uniform.
- Less plugging of perforated aeration floors or ducts occurs.
- Heat moves out of the top of the storage, permitting warm grain to be added to the storage without pulling heat through previously cooled grain.
- Aeration cycle progress is monitored by measuring grain temperature at the top surface.

In a **negative pressure system**

- Aeration cycle progress is monitored by measuring the exhaust air temperature. However, nonuniform airflow (e.g. air going around a pocket of foreign material) results in erroneous temperature readings; exhaust air temperature is only an indirect indicator of grain conditions.
- Less potential for condensation under the bin roof exists because warm air expels directly out through the fan.
- Greater potential for bin roof damage exists due to roof vent screens being blocked by frost, resulting in a vacuum in the storage.
- A 50% larger cross-sectional area in ducts and transitions is needed to hold pressure drop to acceptable limits. This is particularly true in shallow depth flat storage.
- Outdoor grain pile cover tarps can be held down.

Positive pressure systems are recommended due to the ease of determining when the aeration cycle is completed, and to the reduced hazard of the roof being damaged due to frost blocking vent screens.

System Components

This chapter will describe in detail components that make up a dry grain aeration system. Descriptions will include information about the types of components used and explanations about how they interact with other components in the system.

A grain aeration system has the following main components:

- Fans.
- Transitions.
- Supply ducts.
- Manifolds.
- Perforated ducts for air distribution.
- Intakes and exhausts.
- Controls and monitors

One variable that is affected by the design of every component in an aeration system is the static pressure within the system. Static pressure, typically measured in inches of water, is the resistance a fan must overcome to provide the needed airflow through the grain. The resistance to airflow (static pressure) is created by the grain and equipment. The layout of the air distribution system and the components used affect the static pressure. The airflow a fan delivers will decrease as static pressure increases. The following factors affect static pressure:

- As grain depth increases, static pressure increases.
- Transitions, manifolds, ducts, and vents increase static pressure.
- As airflow rates increase, static pressure increases.
- Smaller kernels, fines, and use of grain distributors increase static pressure.

Table 2.1 shows approximate static pressure for several types of grain at selected depths and airflow rates. These values have been adapted from ASAE D272.2, DEC94, and are adjusted to field storage conditions with the multiplier factor given for each grain type. An additional static pressure of 0.5 inches has been added to account for resistance created by ducts, manifolds, transitions, and vents.

Fans

Figure 2.1 shows the types of fans commonly used for aeration:
- Vane-axial.
- Tube-axial.
- Low speed centrifugal (1,750 rpm).
- High speed centrifugal (3,500 rpm).
- In-line centrifugal.

Each fan has specific operating characteristics and applications largely determined by static pressure. Fans usually are driven through a direct connection to the fan motor, but can be driven by V-belts.

Axial-flow fans pass air through the fan in a direction nearly parallel to the fan's axis. Axial fan motors are generally mounted within the airstream. The vane-axial fan has guide vanes that act as air straighteners and reduce energy losses resulting from turbulence of the air leaving the fan blades. Axial-flow fans generally operate at static pressure less than 5 inches. Normally, at static pressure less than 4 inches, these fans give more airflow per unit of energy than centrifugal fans. Tube axial fans are normally small fans used at low static pressure.

Centrifugal fans have a blower wheel (squirrel cage) located in the center of a spiral housing. This wheel has forward, straight, or backward curved blades mounted along its outer edge. Air flows axially into the center of the blower wheel and out radially through the blades. Straight and backward curved blades develop high static pressures and are most commonly used in aeration systems. Low speed centrifugal fans operate at about 1,750 rpm and at static pressure up to about 7 inches. High speed centrifugal fans operate at about 3,500 rpm, and some can develop static pressure in excess of 10 inches. The quantity of airflow from a high speed centrifugal fan at static pressures of 4 to 6 inches normally will be less than a low speed centrifugal fan of the same horsepower and static pressure because of design characteristics. Check manufacturers' data for such comparisons.

In-line centrifugal fans have a centrifugal impeller mounted in the housing of an axial fan. A bell intake funnels air into the impeller. In-line centrifugal fans operate at about 3,500 rpm and can develop static pressure to about 10 inches on 7.5-horsepower or larger fans.

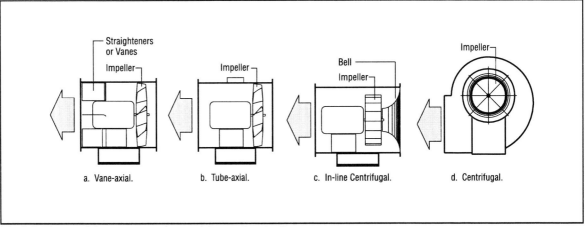

Fig 2.1. Types of fans commonly used for grain aeration.

Table 2.1. Approximate static pressure drop (inches).

Based on ASAE D272.2 DEC94. A multiplier factor is used to convert the pressure drop of clean unpacked grain. An additional static pressure of 0.5" has also been added to account for resistance created by ducts, manifolds, transitions, and vents. When air velocity is greater than 2,500 fpm and for static pressure values above 10" contact an engineer.

Grain type	Grain depth, ft	Airflow rate, cfm/bu							
		0.03	0.05	0.10	0.20	0.25	0.50	0.75	1.00
Barley	10	0.5"	0.5"	0.5"	0.8"	0.9"	1.2"	1.6"	1.8"
Oats	15	0.5"	0.5"	0.8"	1.1"	1.2"	2.1"	3.2"	4.3"
Sunflower (Oil)	20	0.5"	0.5"	1.0"	1.6"	1.9"	3.5"	5.5"	7.8"
(Multiplier = 1.5)	25	0.5"	0.8"	1.3"	2.2"	2.8"	5.6"	8.9"	
	30	0.9"	1.1"	1.7"	3.0"	4.2"	8.1"		
	35	1.0"	1.3"	2.1"	4.0"	5.4"			
	40	1.2"	1.5"	2.7"	5.2"	6.9"			
	50	1.6"	2.1"	4.0"	8.2"				
	60	2.1"	2.9"	5.6"					
	70	2.7"	3.7"	7.5"					
	80	3.5"	4.8"	10.0"					
	90	4.1"	6.0"						
	100	4.9"	7.4"						
Shelled corn	10	0.5"	0.5"	0.5"	0.6"	0.7"	0.8"	1.0"	1.3"
(Multiplier = 1.5)	15	0.5"	0.5"	0.6"	0.8"	0.9"	1.3"	1.9"	2.5"
	20	0.5"	0.5"	0.7"	1.0"	1.2"	2.1"	3.2"	4.6"
	25	0.5"	0.5"	0.9"	1.3"	1.6"	3.1"	5.2"	7.6"
	30	0.5"	0.5"	1.0"	1.7"	2.3"	4.6"	7.7"	
	35	0.7"	0.9"	1.3"	2.2"	3.0"	6.4"		
	40	0.8"	1.0"	1.5"	2.8"	3.9"	8.6"		
	50	1.0"	1.2"	2.1"	4.4"	6.8"			
	60	1.3"	1.6"	3.0"	6.4"	10.1"			
	70	1.5"	2.0"	4.0"	9.1"				
	80	1.8"	2.5"	5.2"					
	90	2.2"	3.1"	6.6"					
	100	2.7"	3.8"	8.3"					
Soybeans	10	0.5"	0.5"	0.5"	0.5"	0.5"	0.7"	0.9"	1.0"
(Multiplier = 1.3)	15	0.5"	0.5"	0.5"	0.7"	0.7"	1.1"	1.4"	1.8"
	20	0.5"	0.5"	0.7"	0.9"	1.0"	1.5"	2.2"	2.9"
	25	0.5"	0.5"	0.8"	1.1"	1.3"	2.2"	3.3"	4.7"
	30	0.5"	0.7"	0.9"	1.4"	1.6"	3.0"	4.8"	7.0"
	35	0.5"	0.8"	1.1"	1.7"	2.1"	4.1"	6.7"	9.8"
	40	0.7"	0.9"	1.3"	2.1"	2.7"	5.4"	9.1"	
	50	0.9"	1.1"	1.7"	3.1"	4.0"	9.8"		
	60	1.0"	1.3"	2.2"	4.4"	5.9"			
	70	1.2"	1.6"	2.9"	5.9"	8.9"			
	80	1.5"	2.0"	3.7"	7.8"				
	90	1.8"	2.4"	4.6"	10.0"				
	100	2.0"	2.9"	5.7"					
Wheat	10	0.5"	0.5"	0.5"	0.9"	1.0"	1.6"	2.1"	2.8"
Grain sorghum	15	0.5"	0.5"	1.0"	1.4"	1.7"	3.0"	4.5"	6.1"
(Multiplier = 1.3)	20	0.5"	0.5"	1.3"	2.2"	2.7"	5.1"	7.9"	
	25	0.5"	1.2"	1.8"	3.2"	4.0"	8.0"		
	30	0.5"	1.5"	1.9"	4.4"	5.5"			
	35	1.3"	1.8"	3.1"	5.9"	8.0"			
	40	1.6"	2.2"	3.8"	7.6"				
	50	2.2"	3.1"	5.8"					
	60	2.9"	4.2"	8.2"					
	70	3.6"	5.6"						
	80	5.1"	7.2"						
	90	6.1"	9.0"						
	100	7.5"							

Fan Selection

Centrifugal fans are quieter than axial fans; use them when lower noise levels are desired. Centrifugal fans are more expensive to purchase than axial fans, but at higher static pressures centrifugal fans are more efficient and may be less expensive to operate.

Select fans based on these criteria:
- Expected static pressure vs. airflow.
- Efficiency, expressed in cfm/watt.
- Noise.

There are many variations in fan design. Always select fans based upon intended use and performance data. Fan data specifies the volumetric airflow rate the fan will deliver, in cfm, for a range of static pressures. Fan data in static pressure vs. cfm is usually provided in tables, but may be in the form of fan performance curves on graphs.

Examples of fan performance data are shown in Table 2.2. For fans under consideration, always obtain data that has been developed under procedures specified by the Air Movement and Control Association, Inc. (AMCA).

The AMCA is the nonprofit trade association of the world's manufacturers of air movement and control equipment--primarily fans, louvers, dampers, and related air systems equipment. AMCA oversees the Certified Ratings Program, which was established to ensure reliable published performance data of fans and related equipment. Fans are not required to be tested; therefore, some fans are marketed with performance data that has not been evaluated by AMCA or another reputable independent laboratory. An AMCA Certified Ratings Seal as seen in Figure 2.2, indicates the following:

Table 2.2. Example fan performance data.

Data presented is a composite from several manufacturers. Consult a comparable table or performance curve for the actual fans being compared.

Power, hp	Diameter, in	Speed, RPM	1"	2"	3"	4"	5"	6"	7"	8"	9"	10"
Axial												
0.50	12	3,450	1,500	630								
0.75	12	3,450	1,700	750								
1	14	3,450	2,880	1,050								
1.5	12	3,450	2,200	950								
1.5	14	3,450	2,800	1,300								
1.5	16	3,450	3,500	2,400	1,300							
1.5	18	3,450	4,350	3,000	1,400							
3	18	3,500	5,700	4,600	2,650	1,400						
5	24	3,500	10,500	9,000	7,000	4,600	2,900					
7.5	24	3,500	12,500	11,100	9,450	6,550	3,900					
10	26	3,500	15,500	14,000	12,250	9,500	5,800	3,400				
Low Speed Centrifugal												
3	22	1,750	4,580	4,230	3,820	3,350	2,550					
5	24	1,750	7,800	7,000	6,250	5,550	4,600	3,300				
7.5	24	1,750	10,550	9,750	8,950	8,000	7,400	6,100				
10	27	1,750	13,300	12,400	11,550	10,500	9,550	8,500	7,300			
High Speed Centrifugal												
3	13	3,500	-	2,950	-	2,550	-	2,120	-	1,650	-	1,000
5	15	3,500	-	4,350	-	3,850	-	3,200	-	2,200	-	1,800
7.5	16	3,500	-	5,700	-	5,100	-	4,500	-	3,800	-	2,900
10	18	3,500	-	6,800	-	6,300	-	5,750	-	5,100	-	4,450
In-line Centrifugal												
3	18	3,500	3,800	3,600	3,400	3,000	2,500	1,900				
5	24	3,500	5,500	5,000	4,400	4,100	3,900	3,600	2,800	1,800		
7.5	28	3,500	6,200	6,000	5,700	5,500	5,200	4,800	4,500	4,000	3,500	3,000
10	28	3,500	7,700	7,300	6,800	6,500	6,300	6,000	5,400	5,100	4,800	4,400

- The manufacturer's published ratings are based on tests conforming to the appropriate AMCA test standard conducted in an AMCA Registered Laboratory or in the AMCA Laboratory.
- The test results have been reviewed and approved by the AMCA staff.
- A fan selected by AMCA has passed a pre-certification test check in the AMCA Laboratory.
- Catalogs containing Certified Ratings have been submitted to the AMCA staff for approval before publication.
- Each licensed product line is check-tested on a periodic basis in the AMCA Laboratory.

Example 2.1: Fan Selection

Select a fan that provides about 2,850 cfm at an expected operating static pressure of 3.2". Table 2.3 lists four fans from Table 2.2 that meet the specifications.

Table 2.3. Fan comparison for Example 2.1

Fan	Static Pressure	cfm[a]
3 hp axial	3"	2,650
3 hp low speed centrifugal	3"	3,820
3 hp high speed centrifugal	3"	2,750
3 hp in-line centrifugal	3"	3,400

[a] Fans selected are based on Table 2.2
Design airflow rate; actual airflow delivered will be less.

The 3-horsepower axial fan is likely the least expensive to purchase but will provide slightly less airflow than required. The 3-horsepower, low speed centrifugal fan would provide the most airflow of the three types of centrifugal fans but would likely be more expensive to purchase. The axial or high speed centrifugal fans would be adequate provided a longer aeration cycle time is acceptable. A conservative design, however, would anticipate the difference between design airflow and delivered airflow and select a fan that could accomplish the job in less-than-ideal conditions.

Fan Power

Required fan motor horsepower is based on the volume of air to be moved, static pressure, and fan efficiency. Once the required airflow and estimated static pressure are determined, estimate the power requirement of the fan motor with equation 2.1.

Figure 2.2. AMCA Certified Ratings Seal.

Motor horsepower is a convenient, but inadequate description for a fan. Regardless of type, fans are best described by specific fan performance data.

Some fan motors show a dual rating, such as 5 to 7 horsepower. Axial and in-line centrifugal fans are air-over motor designs that provide air cooling for the motor and allow it to provide more power without overheating. For example, a 5- to 7-horsepower rated motor could have an output of 5 horsepower without airflow over it, but could provide 7 horsepower with airflow over the motor. The motor load factor may permit a fan to load the motor beyond the nameplate horsepower value. For example, a 5-horsepower motor with a load factor of 1.4 could be loaded to 7 horsepower.

Equation 2.1 Fan Motor Horsepower

$$EFMHP = \frac{(Q)(SP)}{6,350(SE)}$$

Where:
EFMHP = Estimated fan motor horsepower, hp
SP = Static pressure, inches
Q = Total required airflow, cfm
SE = Static efficiency, decimal. Ratio of the theoretical power needed to move the air against the static pressure to the actual power of the motor. Typical values range from 0.40 for lower horsepower fans to 0.70 for higher horsepower fans.

Figure 2.3 compares the relative performance of a 3-horsepower vane-axial, a 3-horsepower low speed centrifugal, a 3-horsepower in-line centrifugal, and a 3-horsepower high speed centrifugal fan. Figure 2.4 shows similar comparisons for 10-horsepower fans. Relative performance is compared between three 18-inch diameter 3-horsepower axial fans in Figure 2.5 and between three, low-speed, 5-horsepower centrifugal fans in Figure 2.6. Again, always obtain manufacturers' data for the specific fans being considered.

Fan Efficiency

If possible, compare fans based on cfm/watt as well as airflow delivery. Electrical efficiency is measured in cfm/watt much like fuel efficiency is measured in miles per gallon. Aeration costs are directly related to cfm/watt.

Fan electrical efficiency rated as cfm/watt
- Can vary widely between fans at the same static pressure.
- Always goes down as static pressure goes up. (Keep static pressure as low as practical.)
- Is not necessarily better for larger horsepower motors.

Although electrical efficiency data for fans is seldom published, it should be available from the manufacturer. Select the fan with the highest electrical efficiency at the expected operating static pressure.

Multiple Fans

Three types of multiple fan systems used in aeration systems are series, parallel, and multiple duct. Use a multiple fan system when more airflow or static pressure is needed than can be supplied by a single fan. It is generally less expensive to purchase a single fan for the desired airflow and static pressure situation. On the other hand, a multiple duct system may be easier and simpler to set up with multiple fans than a single fan with a manifold.

Axial fans can be connected in *series*, also referred to as in tandem, Figure 2.7. This installation approximately doubles the static pressure that can be developed, which is important when an axial fan is used for static pressure applications higher than a single axial fan can operate efficiently. For example, Table 2.2 shows that a 5-horsepower, 24-inch diameter axial fan operating at 3,500 rpm is expected to have an airflow of 7,000 cfm at 3 inches of static pressure. Two of these fans in series would provide an airflow of about 7,000 cfm at 6 inches static pressure. In comparison, a 10-horsepower, 27-inch

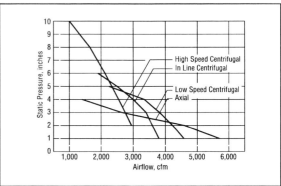

Figure 2.3. Performance curves of four 3-hp rated aeration fans.

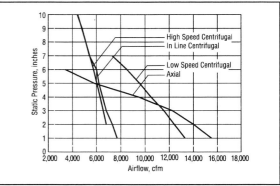

Figure 2.4. Performance curves of four 10-hp rated aeration fans.

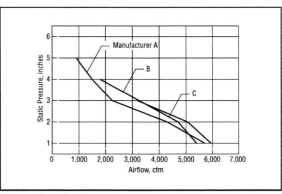

Figure 2.5. Performance curves of three manufacturers' 3-hp axial-flow fans, with 18" diameter.

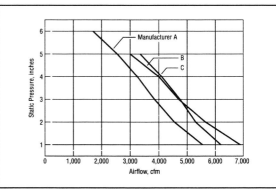

Figure 2.6. Performance curves of three manufacturers' 5-hp, low-speed centrifugal fans.

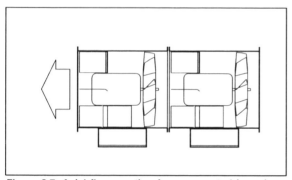

Figure 2.7. Axial flow aeration fans connected in series.
Static pressure is additive; airflow remains constant.

diameter low speed centrifugal fan operating at 1,750 rpm would provide about 8,500 cfm at 6 inches of static pressure.

Fans also can be connected in *parallel*, Figure 2.8. This installation increases the quantity of airflow, but not as much as might be expected. In general, doubling the airflow requires five times the fan horsepower.

For example, a 5-horsepower centrifugal fan operating at 1,750 rpm moves about 5,100 cfm through 17.5 feet of wheat in a 30-foot diameter bin operating at 4.2 inches of static pressure. Two of these fans in parallel move about 6,940 cfm through the wheat operating at 6 inches of static pressure. Doubling the horsepower increases airflow by 36%. If static pressure exceeds the capability of the fans, there is no benefit from fans in parallel. Expected airflow and static pressure need to be computed or determined from design tables.

Air Distribution Systems

The design of a grain aeration system depends on the following criteria:
- Type of grain stored.
- Shape of the grain surface.
- Shape of the grain storage structure.
- In-floor or on-floor ducts.
- New construction or remodelling.

A variety of air distribution systems exist. Components for these systems include
- Perforated ducts, floors, or pads as designed.
- Non-perforated ducts and floors, as needed.
- Transitions, supply ducts, and manifolds.

Airflow Path Ratio Design

The main design criteria to obtain acceptable airflow uniformity for aeration systems is the airflow path ratio. The airflow path ratio is the ratio between the longest and shortest airflow path. In most cases the airflow path ratio should not exceed 1.5:1, Figures 2.9 and 2.10.

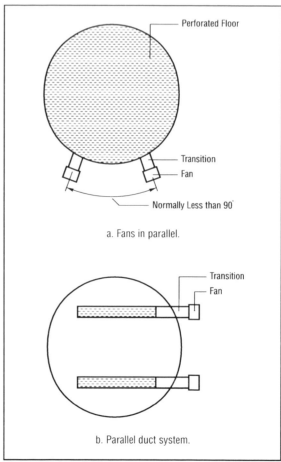

a. Fans in parallel.

b. Parallel duct system.

Figure 2.8. Parallel fan systems

Using the 1.5:1 airflow path ratio will create nonuniform airflow, therefore causing a variation in stored grain cooling or warming times, but the variation is tolerable. Airflow paths and their corresponding estimated cooling and warming times are shown for a design airflow rate of 0.10 cfm/bu, Figure 2.10 and Table 2.4. The average airflow path will be at the location in the grain where the design airflow rate is moving through the grain.

A fan on an aeration system with a design airflow rate of 0.10 cfm/bu would need to operate for 60 hours longer to aerate the slowest cooling grain after the fastest cooling grain has been cooled. The average of the aeration times has been used in Table 1.1.

Air Distribution Components

Select or design a distribution system that permits periodic cleaning of the enclosed space under any perforated surface to remove foreign material, grain, and insect or rodent debris. Accumulation leads to blocked airflow, reduced fan efficiency, higher operating costs, and an environment conducive to insect and fungi infestations.

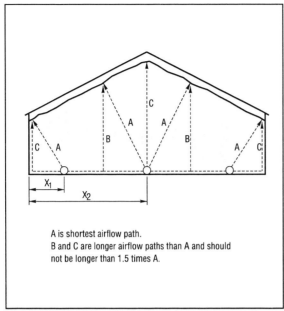

Figure 2.9. Duct location's impact on airflow path ratio in flat storage.

All air distribution components must be structurally adequate to support the load imposed by the grain. Do not use duct materials that were not specifically designed for grain aeration. The maximum grain load can be estimated by multiplying the grain depth by the grain bulk density, lb/cu ft (ASAE EP433 & EP545). Consult an engineer knowledgeable about grain structures for design assistance.

Duct Systems

Correctly designed aeration duct systems provide satisfactory airflow uniformity. They are adaptable to bins and flat storages. Duct systems meet the unique requirements of flat storages where the width or length of the grain mass substantially exceeds grain depth. Size ducts for the airflow rate of the fan selected, which in most cases will be different than the design airflow rate.

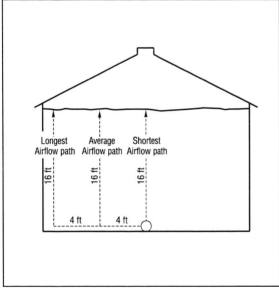

Figure 2.10. Airflow paths for a bin of stored grain.

Table 2.4. Airflow rates and aeration cycle times.

Using airflow path ratio design method. Refer to Figure 2.10.

	Longest Path	Average Path	Shortest Path
Airflow Ratio:	1.5	1.25	1.0
Path Length, ft:	24	20	16
Airflow Rate, cfm/bu:	0.083	0.10	0.125
Aeration Time, Hrs	180	150	120

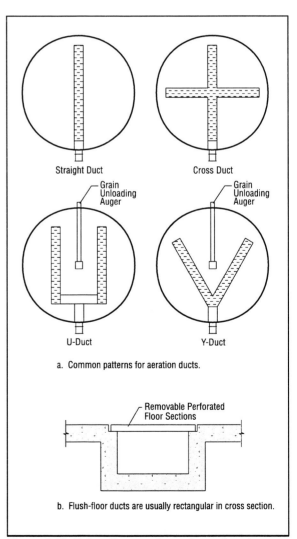

a. Common patterns for aeration ducts.

b. Flush-floor ducts are usually rectangular in cross section.

Figure 2.11. Below-floor perforated aeration ducts.

Chapter 2

System Components

Air Distribution
Systems

Air Distribution
Components

Duct Systems

Below-floor ducts

Aeration ducts below a concrete floor result in an obstacle-free floor, Figure 2.11. Recessed ducts must be adequately reinforced or bridged to permit vehicle traffic when necessary. Size ducts to minimize waste of the removable perforated floor section material. Common duct widths range from about 20 to 36 inches. Supports may be needed to provide the required structural strength.

For level-filled storages, space ducts at intervals equal to the grain depth, Figure 2.12. Use Equations 2.2 and 2.3 to calculate the number of duct locations needed and space between ducts.

Above-floor ducts

Aeration ducts installed on the floor are relatively low cost and can be adapted to almost any grain storage structure, Figure 2.13. However, the ducts can be obstructions during unloading

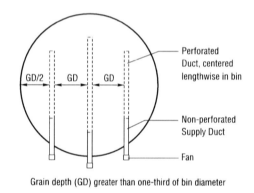

A. Grain depth greater than bin diameter (BD)

Grain Depth (GD) greater than one-half of bin diameter

Grain depth (GD) greater than one-third of bin diameter

Figure 2.12. Recommended duct spacing.

Equation 2.2 Number of Ducts

$$ND = \frac{W \text{ or } BD}{GD} \quad \begin{array}{l}\text{Round to the next}\\ \text{larger whole number.}\end{array}$$

Where: ND = Number of ducts
BD = Bin diameter, ft
W = Building width, ft
GD = Grain depth, ft

Equation 2.3 Duct Spacing Level Fill

$$SL = \frac{W \text{ or } BD}{ND}$$

Where: SL = Duct spacing level fill, ft
ND = Number of ducts
BD = Bin diameter, ft
W = Building width, ft

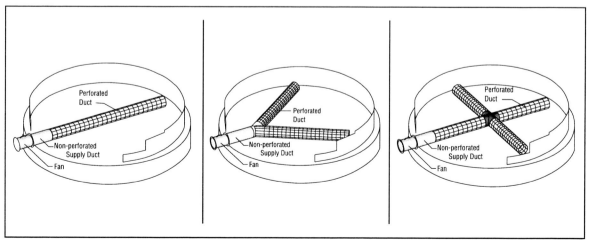

Figure 2.13. Above-floor perforated aeration ducts.

(and sometimes filling) and can become damaged as they are removed, stored, and reinstalled each year. Above-floor ducts are readily adaptable to an emergency need and may be justified for only a few years storage. Correctly designed above-floor duct systems are well suited for dry grain maintenance using low aeration rates. Above-floor duct systems are often used in grain piles with both level and peaked grain surfaces.

Commercial round and semicircular ducts are available in diameters up to 36 and 48 inches respectively, and in lengths up to 20 feet. Although ducts can be field cut to any length, it is best to order the length needed to the closest 1 or 2 feet. Install ducts so no openings in the ducts permit grain to enter and block airflow through the duct.

Duct Sizing

Duct diameters must be large enough to properly carry the airflow without exceeding the maximum design duct air velocity. High air velocities increase the fan power required and increase the variation in airflow exiting from the duct. Size perforated ducts in bins for a maximum duct air velocity of 2,000 fpm. In flat storages, size for a maximum air velocity of 1,500 fpm, to improve the airflow uniformity.

The velocities are different in bins and flat storages due to the difference in ratios between typical duct lengths and grain depths. A typical duct length in a bin is 20 feet with a 20-foot grain depth. A typical flat storage duct length is 70 feet with a grain depth of only 10 feet. At an airflow rate of 0.2 cfm/bu, the static pressure for moving air through 20 feet of wheat is 1.7 inches, but is only 0.4 inches for a 10-foot depth. A difference in static pressure of 0.5 in the duct has a larger impact in the shallow depth associated with flat storage, 0.5/0.4 = 1.25, than in the 20-foot depth typical of a bin, 0.5/1.7=0.29.

Use positive pressure systems for ducts in flat storages for improved airflow uniformity. Negative pressure systems may be used with outdoor piles when the fan helps to hold a tarp cover on the grain. Use a maximum duct air velocity of 1,000 fpm for negative pressure systems.

Airflow uniformity can be improved by controlling the amount of perforated duct surface area. Airflow distribution uniformity is best when the effective open area in the duct surface is 90 to 100% of the duct cross-sectional area. Because grain typically blocks some of the open area, the actual open area needs to be about three times the duct cross-sectional area.

The maximum allowable perforated duct surface air velocity is 30 fpm for ducts with at least 10% of the surface area open (20 fpm maximum for negative pressure systems). There is very little airflow resistance through the surface of commercial ducts designed for grain aeration if adequate surface area is maintained.

There may be differences in the quantity of airflow exhausting from openings along the length of the duct. These differences result in different aeration cycle times along the length of the duct. The variation in airflow uniformity increases as the duct length increases. Generally, limit duct length to less than 100 feet.

The duct cross-sectional area, airflow capacity at selected air velocities, perforated surface

Table 2.5. Round duct characteristics.

Diameter, in	Cross-sectional area, sq ft	Airflow, cfm		Surface area[a] sq ft/ft of length	Maximum airflow through perforation, cfm/ft of length
		Flat storage	Round bins		
6	0.2	300	400	1.26	38
8	0.3	525	700	1.68	50
10	0.5	825	1,100	2.09	63
12	0.8	1,200	1,600	2.51	75
14	1.1	1,600	2,150	2.93	88
15	1.2	1,850	2,450	3.14	94
16	1.4	2,100	2,800	3.35	101
18	1.8	2,650	3,550	3.77	113
21	2.4	3,600	4,800	4.40	132
24	3.2	4,700	6,300	5.03	151
30	4.9	7,350	9,800	6.28	188
36	7.1	10,600	14,150	7.54	226

[a] Surface area reduced by 80% to compensate for restriction caused by the floor.

Table 2.6. Semicircular duct characteristics.

Diameter, in	Cross-sectional area, sq ft	Airflow, cfm		Surface area sq ft/ft of length	Maximum airflow through perforation, cfm/ft of length
		Flat storage	Round bins		
20	1.1	1,650	2,200	2.62	79
25.5	1.8	2,650	3,550	3.33	100
30	2.5	3,700	4,900	3.93	118
36	3.5	5,300	7,100	4.71	141
42	4.8	7,200	9,600	5.50	165
48	6.3	9,450	12,600	6.28	188

Table 2.7. Surface open area of aeration ducts.

Duct type	Open area, %	Comments
Perforated steel aeration duct	greater than 10	Little resistance to airflow.
Perforated plastic aeration duct	2 to 3	Increases resistance to airflow by about 10%. The ducts are perforated at the base of the corrugations and wrapped with diffusion screen to prevent kernels from blocking the openings.
Perforated drainage tile	less than 1	May significantly increase resistance to airflow. Effective surface area should not be less than 90% of the duct cross-sectional area.

area and airflow discharge per linear foot of duct for some round and semicircular ducts, are shown in Tables 2.5 and 2.6. Table 2.7 compares characteristics of different types of ducts available for aerating grain.

Air velocity depends on the duct cross-sectional area (sq ft) and duct airflow rate (cfm). When two of the three factors are known, the other factor can be calculated with one of the following formulas:

Equation 2.4 Airflow

$$Q = (A)(V)$$

Equation 2.5 Duct Area

$$A = \frac{Q}{V}$$

Equation 2.6 Air Velocity

$$V = \frac{Q}{A}$$

Where: Q = Total airflow, cfm
A = Duct cross-sectional area, sq ft
V = Air velocity, fpm

Example 2.2: Minimum Cross-sectional Area
Determine the minimum duct cross-sectional area needed to move an airflow of 4,500 cfm. The design air velocity is 1,500 fpm. Using Equation 2.5:
    A = Q / V
    A = 4,500 cfm / 1,500 fpm
       = 3 sq ft

Duct surface air velocity is defined as the velocity of air exiting the perforated duct surface. The required perforated surface area for a specified surface air velocity is calculated by:

Equation 2.7 Perforated Surface Area

$$AS = \frac{Q}{VS}$$

Where: VS = Surface air velocity, fpm
Q = Total airflow, cfm
AS = Perforated surface area, sq ft

The surface area for a duct is the surface area per linear foot times the duct length. Only 80% of the total surface area of round ducts is considered usable because the duct sits on the floor. All of the surface area is usable with semicircular ducts.

$$AS = \pi(D)(DL)(.8) \quad \text{(round ducts)}$$

$$AS = \frac{\pi(D)(DL)}{2} \quad \text{(semicircular ducts)}$$

Where: D = Duct diameter, ft
DL = Duct length, ft
π = 3.1416 (pi)

Chapter 2

System Components

Air Distribution
Components

Perforated Floor

Aeration Pads

Perforated Floor

A perforated floor, Figure 2.14, in a typical 16- to 20-foot deep grain bin is not necessary for satisfactory aeration, but is essential if high airflow rates are selected for drying, holding wet grain prior to drying, or cooling hot grain directly from a dryer. Systems designed for these purposes can be used for aeration but an aeration system cannot be used for drying. Initial costs of such systems are higher, but they provide flexibility in using the bin, as well as optimum operating times as determined by weather or non-peak hour electrical rates offered by the local utility company.

Aeration Pads

Aeration pads are square perforated sections, commonly used in round bins, Figure 2.15. Install the pad over a recessed floor constructed below the concrete bin floor to form an air plenum; the pad surface is flush with the bin floor. A properly sized air supply duct (sized to keep air velocity less than 2,500 fpm) is formed in the concrete floor from the plenum to the edge of the bin. The recessed section of floor is normally the same depth as the supply duct and the same dimensions as

---

Example 2.3: Pad sizing
Determine the minimum pad size for a 30' diameter bin with an 18' grain depth (10,220 bu) aerated at 0.20 cfm/bu.

Using Equation 1.1 and Equation 2.7 with the maximum surface air velocity of 30 fpm

$$Q = 10,220 \text{ bu} \times 0.20 \text{ cfm/bu}$$
$$= 2,044 \text{ cfm}$$

$$AS = \frac{2,044 \text{ cfm}}{30 \text{ fpm}}$$

$$= 68 \text{ sq ft}$$

Minimum pad size based on airflow is 9' x 9' ($\sqrt{68}$ = 8.25, so use next larger size). Use equation 2.8 to determine the minimum pad length.

Equation 2.8 Aeration Pad Length
Minimum pad length for airflow uniformity is calculated by:

$$PL = BD-GD$$

Where: PL = Pad length, ft
         BD = Bin diameter, ft
         GD = Grain depth, ft

$$= 30'-18'$$
$$= 12'$$

The recommended minimum pad size would be 12' x 12'.

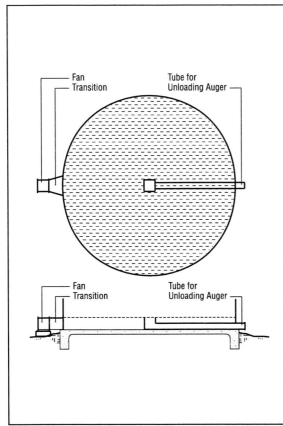

Figure 2.14. Perforated floor.

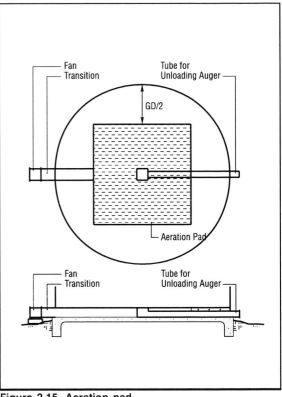

Figure 2.15. Aeration pad.

Chapter 2

System Components

Air Distribution
Components

Channel-lock
Perforated Flooring

Transitions, Supply
Ducts, and Manifolds

the aeration pad. Properly selected and installed aeration pads generally have more uniform airflow distribution than duct systems.

Use the two following criteria to size an aeration pad for a particular setup, then use the larger pad size indicated:

- Calculate pad size based on a maximum surface air velocity of 30 feet per minute.
- Calculate pad size based on the difference between bin diameter and grain depth.

Locate the pad with a maximum distance between pad and bin wall of one half the grain depth to maintain the design airflow path ratio at 1.5:1.

Channel-lock Perforated Flooring

Channel-lock perforated flooring can be installed directly on concrete bin floors. The approximately ¾-inch high air channels are laid perpendicularly over properly sized air supply ducts built below the bin floor, Figure 2.16.

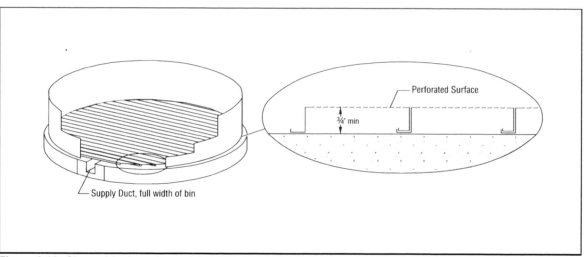

Figure 2.16. Channel-lock perforated flooring on concrete bin floor.

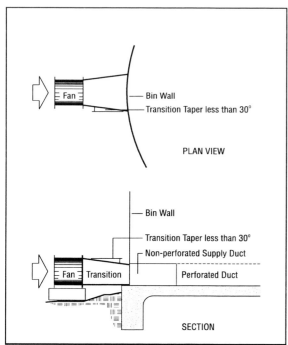

Figure 2.17. Transition taper and supply duct for air distribution systems.

These installations are generally less expensive than raised perforated floors and provide uniform airflow at aeration rates of 0.10 cfm/bu or less. Size supply ducts and areas under the flooring to limit air velocity to 2,000 fpm (because this is a perforated duct rather than a solid). The space under the channels collects foreign material and must be cleaned annually.

Transitions, Supply Ducts, and Manifolds

The fan is connected to the air distribution system by a transition and/or supply duct. The function of either of these is to transfer air from the fan to the air distribution system with a minimum increase in static pressure. In general, avoid sudden changes in air direction, constrictions, and rough interior duct surfaces. Match the entrance of the transition to the size and shape of the fan outlet as closely as possible. The transition taper should be less than 30 degrees, Figure 2.17.

Non-perforated supply ducts convey air to the perforated duct air distribution system, Figure 2.17. The maximum design supply duct air velocity is 2,500 fpm. For ducts with very rough surfaces, such as some corrugated ducts, reduce velocity to limit friction loss of the airflow caused by the roughness of the duct.

Selecting a fan for each perforated duct allows independent control and management of each duct. For example, if a hot spot is discovered near one of the ducts, that duct fan can be operated to cool the hot spot. However, be aware that some of the air will come out of nonoperating fans unless backdraft dampers are installed.

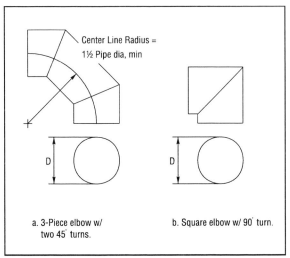

Figure 2.18. Two types of elbows for making turns in supply ducts.

> **Example 2.4: Aeration supply duct design.**
> Size an air supply duct for an airflow rate of 2,044 cfm. Using Equation 2.5, with the maximum air velocity of 2,500 fpm:
> $$A = 2,044 \text{ cfm} / 2,500 \text{ fpm}$$
> $$= 0.8 \text{ sq ft}$$
>
> Using a depth of 8" for ease of construction, results in a supply duct with a width of 15" (0.8 sq ft / (8"/12) = 1.2' = 14.4").

A manifold connects two or more ducts to a fan, reducing the number of fans required and simplifying the aeration control system. A manifolded duct system distributes air to desired perforated ducts with slide gates or dampers.

When a fan is selected for more than one perforated duct connected into a manifold system, specify the fan airflow and static pressure as follows:

- **Total fan airflow**: add the airflow for each duct.
- **Design static pressure**: use the duct developing the maximum static pressure.

Design the manifold to provide adequate duct cross-sectional areas to carry airflow and have gradual changes in airflow direction. The area of the air supply duct should at least equal the sum of the cross-sectional areas of the perforated ducts connected to the air supply duct. Design air velocity will be approximately equal in all ducts that are connected to a single fan, so ducts of equal cross-sectional area will have equal airflow assuming equal resistance. The resistance through a three-piece elbow, Figure 2.18a, is only about one-fourth of the resistance through a square elbow, Figure 2.18b. Use elbows with a centerline radius equal to at least 1½ duct diameters, Figure 2.18a. The length of the supply duct and manifold should be minimized to limit friction loss. Example 4.4 includes the design of a manifold.

Vertical Aerators

Vertical aerators have limited application. They are useful for localized cooling, e.g. a small pocket of hot grain. Most vertical aerators use a 4- or 6-inch diameter tube to form the air supply and perforated pipe, Figure 2.19. Some have a helix welded onto the outer surface so the pipe can be screwed into the grain.

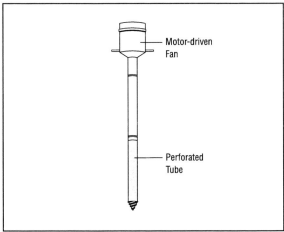

Figure 2.19. Vertical aerators.

Hopper Bottom Bins

In hopper bottom bins, a portion of the hopper can be perforated to serve as the aeration duct, or ducts can be mounted on the hopper surface, Figure 2.20. Adequate surface area and duct cross-sectional area must be provided. Provide 1 square foot of perforated surface area for each 30 cfm and 1 square foot of duct cross-sectional area for each 2,000 cfm. Multiple ducts provide more uniform airflow distribution through the grain.

The capacity of grain in the hopper portion can be estimated from Table 2.8. The table also gives the maximum length of duct that may be placed on one side of the hopper for various bin diameters and for two hopper slopes.

In Figure 2.21, the air makes a turn when it leaves the fan and transition and enters the duct, creating high static pressure. For this reason, use a non-perforated duct next to the fan for a distance equal to twice the duct diameter. A surface within one fan diameter of the fan that causes a sharp change in airflow direction greatly increases static pressure and reduces airflow from the fan.

Another airflow distribution option for a hopper bottom bin uses a perforated plenum supported in the middle of the hopper. This option permits more perforated area, and, because a supply duct carries the airflow in a straight path, the airflow near the fan is less obstructed than with some distribution systems.

Use Table 2.1 to determine the expected operating static pressure and fan charts similar to Table 2.2 for selecting the appropriate fan.

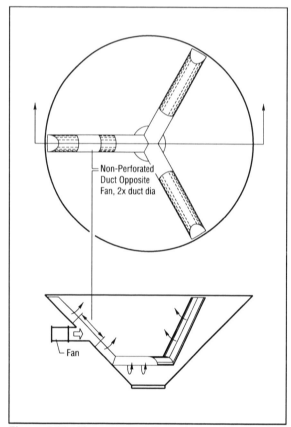

Figure 2.21. Air movement through transition and ducts in hopper bottom bin.

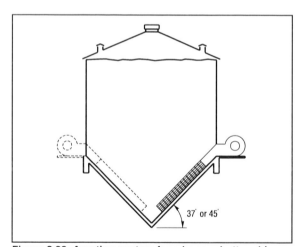

Figure 2.20. Aeration system for a hopper bottom bin.

Table 2.8. Capacities and duct lengths for hopper bottom bins.

See Figure A.1 for shape formulas.

Bin diameter, ft	45° cones, 1 : 1 slope		37° cones, 3 : 4 slope	
	Capacity, bu	Maximum length of duct, ft	Capacity, bu	Maximum length of duct, ft
14	280	9	210	8
18	610	12	450	10
21	960	14	720	12
24	1,440	16	1,080	14
27	2,060	18	1,540	16
30	2,820	20	2,120	18
36	4,880	24	3,660	21
42	7,750	28	5,810	25

Silos

Both concrete stave and sealed (oxygen limiting) upright silos have been converted and used successfully to store dry grain.

Check the structural integrity of the silo because the loads imposed by dry grain exceed those of silage. Contact the manufacturer or erector of the structure, or a structural engineer for the evaluation. Install a grain sump in the middle of the silo floor to remove the grain uniformly. Do not remove grain from one side of the silo or structural damage can occur.

Aerate dry grain stored in a silo to prevent moisture migration and possible serious grain spoilage. Although perforated floors can be installed, a duct system is adequate and is considerably less expensive. Install the perforated duct to at least the middle of the silo.

Install perforated ducts in sealed silos by properly supporting perforated metal flooring over the silo unloader trench. Install the grain unloading auger into the trench to an unloading sump located in the perforated metal in the center of the silo, Figure 2.22.

A round aeration duct can be installed in a conventional silo, Figure 2.23a. Remove the bottom silo door and replace it with ¾-inch plywood. Insert the round steel aeration duct through the lower part of the plywood and support it on the silo door frame, 2.23b. The open end of the perforated duct is located in the center of the silo on a raised pad, such as two concrete blocks. The inclined aeration duct must be supported to prevent the duct from collapsing, Figure 2.23c.

Remove grain from the silo by removing the aeration fan and inserting an unloading auger into the duct.

A positive pressure aeration system is recommended for conventional silos to prevent pulling water through wall cracks if the fan is operating during wet weather.

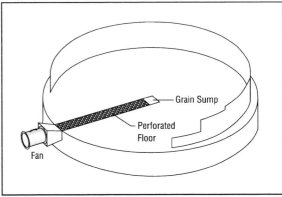

Figure 2.22. Aeration duct arrangement for a sealed silo used to store dry grain.

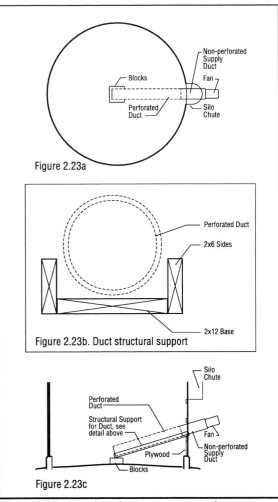

Figure 2.23a

Figure 2.23b. Duct structural support

Figure 2.23c

Figure 2.23. Aeration duct arrangement in a conventional silo converted to store dry grain.

Intakes and Exhausts

Bin air exhaust velocity (positive pressure systems) and entrance velocity (negative pressure systems) should not exceed 1,000 fpm. Values over 1,000 fpm result in increased static pressure, requiring larger fans and increased system cost. Also, increased static pressure increases the structural load on the roof.

Design minimum vent area based on a maximum air velocity of 1,000 fpm. Examples of exhaust or entrance areas are

- Eave space between the roof and walls.
- Space between the grain filling hatch and cover; fill hatch, if left open.
- Roof access door, if left open.
- Commercial roof vents.

Minimum vent area is calculated by using Equation 2.9, with velocity of 1,000 fpm.

This eave space is adequate for an airflow up to 2,400 cfm. Install additional free air vent area, such as a roof vent, for larger airflows.

Roof vents are normally equally spaced around the bin. If there is more than 0.125 inch of static pressure difference between the head space of a storage bin and the outside, when the fan is running with grain in the bin, the air entrance or exhaust area is inadequate, Figure 2.24.

Wire screens protect inlets and outlets from rodents and birds. Avoid screens with small openings that plug with chaff or frost. A ½-inch hardware screen is satisfactory.

Roof vents can become totally plugged due to frost. On a negative pressure system, the roof may collapse if the fan horsepower and air seal are sufficient. Use a static pressure sensor to protect such a system or avoid operating the fan when the outside temperature is near freezing with high relative humidity—above about 80%.

Moisture condensation on the underside of the bin roof may be a problem with positive pressure aeration systems when large temperature differences exist between the grain and outside air. To minimize condensation evaluate the management plan and reduce the grain temperature to within 10 to 15 degrees of the outside air temperature. If a change in management is not possible, install additional vents or roof exhaust fans.

Equation 2.9 Vent Area

$$AV = \frac{Q}{1000}$$

Where: AV = Minimum vent area, sq ft
         Q = Total airflow, cfm

Equation 2.10 Bin Eave Opening

$$EO = \frac{\pi(BD)(EW)}{12}$$

Where: BD = Bin diameter, ft
         EO = Eave opening, sq ft
        EW = Width of eave opening, in
         π = 3.1416 (pi)

Example 2.5: Eave Area
Determine the area of the eave space of a 36' diameter bin with a 0.25" wide eave opening. Using Equation 2.10:

$$EO = \frac{\pi(36)(0.25)}{12}$$

$$= 2.4 \text{ sq ft}$$

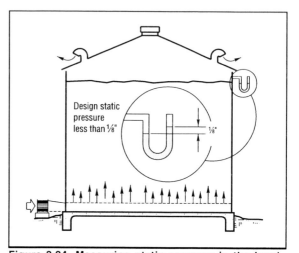

Figure 2.24. Measuring static pressure in the head space of a bin.

Controls and Grain Monitoring Equipment

Equipment that monitors and controls the aeration process include:

- Simple on/off motor switches.
- Temperature and humidity controls.
- Sophisticated programmable controllers with memory.

This equipment is no substitute for proper system operation and stored grain management. Regular inspections of grain and equipment are still required. Proper use of controls and monitors only make stored grain management easier.

Controllers automatically turn aeration fans on and off according to the time of day or predetermined weather conditions. Label any fan that has an automatic controller so personnel are warned to shut off the power before working on the system or entering the storage. Provide adequate vent area if controllers are used on the fans; bins may be structurally damaged if fans operate without adequate vent area.

Controllers need an *on-off-auto* switch ahead of the automatic control circuit to allow the operator to bypass the control and manually operate the fan. Mount all controls in weatherproof boxes. If the fan motors are less than 1.5 hp, line-voltage controls may be used. With larger fans, use a magnetic motor starter. Employ a qualified electrician to install control devices.

Wherever an automatic control is used, connect a running-time meter to record hours of fan operation to estimate when a cooling or warming cycle is completed.

Thermostats, Humidistats

The simplest automatic control system is a single thermostat, Figure 2.25a. This system turns on the fan anytime the outdoor temperature rises or falls below the predetermined temperature setting.

A two thermostat control system, Figure 2.25b, uses the first thermostat to control the upper temperature limit and the second thermostat to control the lower limit. The fan operates only when the outside air is within the preset temperature range. For example, if the first thermostat is set at 50 F and the second thermostat is set at 35 F, the fan runs only when the outside temperature is between 35 F and 50 F.

Figure 2.25c is identical to Figure 2.25b except that a humidistat is connected in series with the thermostats to turn off fans when ambient humidity rises above a predetermined level. An upper limit of about 90% relative humidity is suggested for fall cooling, and about 70% for spring warming. A humidistat is intended to regulate operation during times when excessive grain rewetting could occur, which could lead to moisture damage in the grain. Humidistats lose calibration and need to be re-calibrated about once a year. Many humidistats on the market cannot be set in the 90% range. One with the desirable range may be difficult to locate.

Time Clocks

A 24-hour time clock operates fans during certain hours of the day regardless of temperature or humidity. It is reliable, but will probably require more supervision and manual control during unsuitable weather.

Microprocessors

Microprocessors are programmed to run fans according to climatic conditions, grain conditions and desired management strategy. These units have sensors to determine climatic and sometimes grain conditions.

Microprocessors evaluate pertinent weather data and grain condition factors, then automatically select a more elaborate management strategy than is possible with some simpler controls or with manual on/off operation. A microprocessor can wait until conditions are appropriate and then reliably operate fans automatically. A microprocessor cannot anticipate weather changes or identify why certain grain conditions exist.

Different controls operate on somewhat different management strategies. Be sure that the controls purchased use the strategy desired. Even the most sophisticated control can fail and does not diminish the need for regular system and grain inspection.

The complexity of controls and sensors makes it more difficult to check them for correct operation. Discovering a faulty control may take some time; by then the opportunity to do the aeration job correctly may be lost.

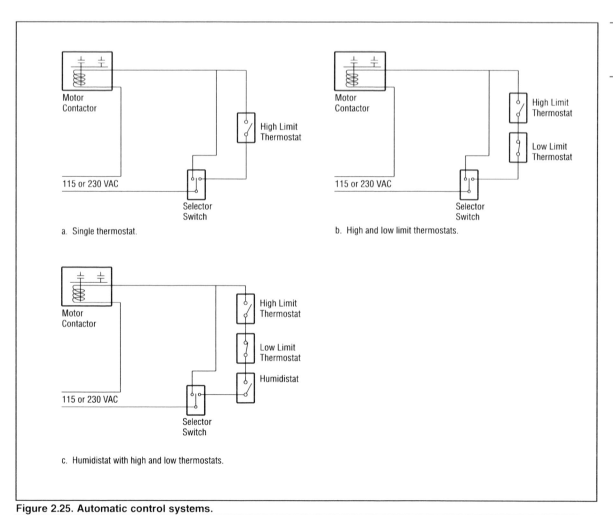

a. Single thermostat.

b. High and low limit thermostats.

c. Humidistat with high and low thermostats.

Figure 2.25. Automatic control systems.

Design Recommendations and Procedures

Chapters one and two have discussed aeration considerations and components. This chapter builds on that information to present design recommendations and procedures for a variety of applications, including cylindrical bins and flat storage. Designs for flat storage include level and peaked storage as well as level storage with sloped sides.

Chapter 3

Design
Recommendations
and Procedures

Basic Design
Principles

Cylindrical Bins

Basic Design Principles

- Use an airflow rate of at least 0.10 cfm/bu in the Midwest for farm-stored shelled corn and soybeans. A higher airflow rate, 0.20 to 0.25 cfm/bu, is used for wheat and other grains harvested when limited cooling time is available.
- Use airflow rates of 0.03 to 0.05 cfm/bu for grain depths exceeding 50 feet.
- Note that doubling the airflow rate increases required fan horsepower about five times.

Transitions, Supply Ducts, and Manifolds

- Design for a transition taper of 30 degrees or less.
- Provide for a maximum allowable supply duct air velocity of 2,500 fpm (1,500 fpm for negative pressure systems).
- Use two 45 degree elbows instead of a single 90 degree elbow.
- Make the elbow radius of curvature at least 1.5 times the duct diameter.

Air Distribution Systems

- Design for a maximum airflow path ratio of 1.5:1: the ratio of the length of the longest airflow path to the length of the shortest airflow path. An airflow path is the distance from the perforated flooring or duct to the grain surface.
- Use positive pressure systems for ducts in flat storages.
- Provide a maximum allowable distribution duct velocity of 2,000 fpm (1,500 fpm for flat storages).
- Design for a perforated surface air velocity of 30 fpm.

Intakes and Exhaust

- Install roof vents equally spaced around the bin.
- Provide 1 square foot per 1,000 cfm.
- Make sure the difference in static pressure between the head space of a storage bin and the outside is one-eighth inch or less. If there is more than one-eighth inch of static pressure difference between the head space of a storage bin and the outside, when the fan is running, the air entrance or exhaust area is inadequate.

Cylindrical Bins

Cylindrical bin aeration systems can be configured in a variety of layouts and duct types. See Chapter 2, **System Components**, for common layouts. Ducts can be built into the floor or placed on top of the floor.

Design criteria for cylindrical bin aeration systems are the following:

- Maximum distance between ducts is grain depth. (**NOTE:** Grain depth is assumed equal to floor-to-eave height.)
- Maximum distance between bin wall and perforated duct or floor pad is one half the grain depth.
- Maximum pad and perforated duct surface air velocity is 30 fpm.
- Maximum air velocity in perforated ducts is 2,000 fpm.
- Maximum air velocity in non-perforated supply ducts is 2,500 fpm.
- Maximum intake or exhaust air velocity is 1,000 fpm.

For aeration pad systems, the minimum side length for the square pad is bin diameter minus grain depth or eave height. However, both minimum size and maximum surface air velocity must be checked to determine proper pad size.

The example design problem beginning on page 28 uses the design criteria presented above in a step-by-step process. The worksheet in Figure 3.1 presents these steps and appropriate information to assist with the design process.

Chapter 3

Design
Recommendations
and Procedures

Worksheet:
Cylindrical Bins

Figure 3.1: Worksheet for Cylindrical Bins

Given Information:
Storage dimensions: _____
Grain type: _____, **Grain depth:** ___ ft
Design airflow rate: _____ cfm/bu

CALCULATE:

Step 1: *Calculating Storage Volume*
• Storage Volume, bu, Figure A.1, Table A.4
• Total Required Airflow, cfm, Equation 1.1
• Static Pressure, Table 2.1

Step 2: *Choosing a Fan*
• Choose a fan from manufacturers' data like Table 2.2; use fan airflow for the remainder of the design.

Step 3: *Choosing the System Type*
• Choose system type; perforated floor, pad, duct

IF PERFORATED FLOOR
Step 4: *Sizing the Intake/Exhaust Area*
• Size intake/exhaust area, sq ft, Equation 2.9

IF PERFORATED PAD
Step 4: *Sizing the Pad*
• Size pad, sq ft, Equation 2.8

Step 5: *Sizing the Non-perforated Duct*
• Size non-perforated supply duct, Equation 2.5

Step 6: *Sizing the Intake/Exhaust Area*
• Size intake/exhaust area, sq ft, Equation 2.9

IF PERFORATED DUCT
Step 4: *Determining the Duct Location*
• Determine duct location that meets 1.5:1 air path ratio design rule.

Step 5: *Calculating the Perforated Area*
• Required perforated area, sq ft, Equation 2.7
• Choose duct layout. (Figure 2.11, 2.12, 2.13)

Step 6: *Sizing the Ducts*
• Size duct, Equation 2.5
• Perforated duct
• Minimum perforated duct length (round duct only), ft, Table 2.5, Equation 3.8
• Non-perforated supply duct

Step 7: *Sizing the Intake/Exhaust Area*
• Size intake/exhaust area, sq ft, Equation 2.9

Chapter 3

Design
Recommendations
and Procedures

Example:
Aeration Pad Design

Example 3.1: Aeration Pad Design

Given Information:
Storage dimensions: ___30 ft_____
Grain type: Shelled Corn , **Grain depth:** 18 ft
Design airflow rate: 0.10 cfm/bu

SOLUTION:

Step 1: *Calculating Storage Volume*
Estimate the amount of grain in the storage, total required airflow and static pressure.
The storage volume from Table A.4 is 10,179 bu.
The total required airflow is calculated using Equation 1.1:
    Total required airflow   = 10,179 bu x 0.10 cfm/bu
                             = 1,018 cfm

The estimated static pressure is determined using Table 2.1:
          Grain depth   = 18'
          Static pressure = 0.7"

Step 2: *Choosing a Fan*
• Choose a fan using fan manufacturers' data like Table 2.2.
• Axial fan, 12" dia., 0.5 hp, 1,500 cfm at a static pressure of 1"

Step 3: *Choosing the System Type*
• An aeration pad system will be designed.

Step 4: *Sizing the Pad*
•Determine the pad and non-perforated supply duct size. Using Equation 2.7, the minimum
  area required to maintain the maximum surface air velocity of 30 fpm for the chosen
  fan can be calculated:
        A   = 1,500 cfm / 30 fpm
           = 50 sq ft

• Minimum pad size based on airflow is 7'x7'.
• Now check minimum pad size for airflow distribution using Equation 2.8:
    Pad side length = 30'-18'
                 = 12'

• The larger of these two sizes, 7'x7' and 12'x12', must be used. The required pad
  size is 12'x12'.

Step 5: *Sizing the Non-Perforated Duct*
• Determine the non-perforated supply duct size using Equation 2.5 with the maximum air
  velocity of 2,500 fpm:
        A   = 1,500 cfm / 2,500 fpm
           = 0.6 sq ft

• Using a depth of 6" for the supply duct results in a width of 15"
    (0.6 sq ft / (6"/12) = 1.2' = 14.4").

Chapter 3

Design
Recommendations
and Procedures

Example:
Aeration Pad Design

Step 6: *Sizing the Intake/Exhaust Area*
• Determine the minimum air exhaust or intake area using Equation 2.9:

Exhaust area   = 1,500 cfm / 1,000 fpm

                    = 1.5 sq ft

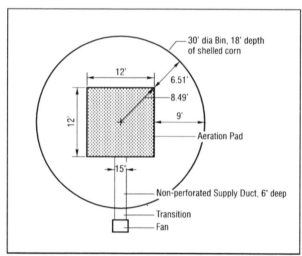

Figure 3.2. Aeration pad design example.

Design
Recommendations
and Procedures

Example:
Aeration System for
Cylindrical Bin

Example 3.2: Aeration System for a Cylindrical Bin

Given Information:
Storage dimensions: _30 ft_
Grain type: _Wheat_ , **Grain depth:** _25_ ft
Design airflow rate: _0.20_ cfm/bu

SOLUTION:

Step 1: *Calculating Storage Volume*
• Estimate the amount of grain in the storage, total required airflow and static pressure. The storage volume is chosen from Table A.4:

Using Table A.4:

Total volume = 1' depth x 25'
= 565 bu/ft x 25'
= 14,125 bu

• The total required airflow is calculated using Equation 1.1:
Total required airflow = 14,125 bu x 0.2 cfm/bu
= 2,825 cfm

• The estimated static pressure is determined using Table 2.1:
Grain depth = 25'
Static pressure = 3.2"

Step 2: *Choosing a Fan*
• Choose a fan using fan manufacturers' data like Table 2.2.

• Axial, 18" dia., 3 hp; 2,650 cfm at a static pressure of 3"

Step 3: *Choosing the System Type*
• A perforated in-floor duct system will be designed.

Step 4: *Determining Duct Location*
• Determine the air distribution system that meets the required 1.5:1 airflow path ratio. Using the design factors presented earlier in this chapter, the maximum distance between the wall and outlet is one half the grain depth, 12.5'. The maximum distance between outlet points is the grain depth, 25'.

• The distance from the wall to the beginning of the perforated duct is determined by centering the perforated duct(s) in the bin lengthwise (see Step 6). When determining duct location, the ratio between the actual and design maximum wall-to-duct distance and the ratio between the actual and design maximum distance between ducts should be similar.

Step 5: *Calculating the Perforated Area*
• Determine the required perforated area using Equation 2.7:
A = 2,650 cfm / 30 fpm
= 88.3 sq ft

• The shape of the duct system can now be determined based on the required perforated area of 88.3 sq ft. A single duct will not meet the required 12.5' maximum

Chapter 3

Design
Recommendations
and Procedures

Example:
Aeration System for
Cylindrical Bin
(continued)

distance between the wall and outlet. A parallel or Y-shaped duct system could provide the required perforated area, meet the spacing criteria and use only a single fan. A Y-shaped duct system will be used in this example, Figure 3.3.

Step 6: *Sizing the Ducts*
- Determine the perforated duct length and width to provide 88.3 sq ft. Use a duct width of 2.5', which is a common width for perforated duct covers. The required length is calculated using the square/rectangle area equation from Figure A.1:

  Total perforated duct length = 88.3 sq ft / 2.5'

                                = 35.3'

- For the Y-shaped duct system each leg would be half of the total length, 35.3'/2 = 17.7'. The duct length is rounded to a whole number, 18'. The required distance from the wall to the start of the perforated duct, Step 4, is determined by centering the perforated ducts in the bin, being careful not to violate the 1.5:1 air path criteria in Step 4. (All points on the bin floor within 12.5' of the perforated duct.)

  Figure 3.3 shows the ratio for the ratio between the actual and design maximum wall-to-duct distance to be 8.5/12.5 = 0.68, and the ratio between the actual and design maximum distance between ducts to be 15.75/25 = 0.63, which are similar.

- Determine the minimum duct cross-sectional area. Each leg will carry 1/2 of the total required airflow, 1,325 cfm. Using Equation 2.5 with a maximum allowable duct air velocity of 2,000 fpm:

  A = 1,325 cfm / 2,000 fpm

       = 0.66 sq ft

- Using a duct width of 2.5' and the square/rectangle area equation from Figure A.1:

  Minimum perforated duct depth   = 0.66 sq ft / 2.5'

                                          = 0.3'

                                          = 4"

- Perforated duct dimensions are 18' x 2.5' x 4".
- Determine the non-perforated supply duct cross-sectional area to limit the maximum allowable non-perforated supply duct air velocity to 2,500 fpm using Equation 2.5:

  A = 2,650 cfm / 2,500 fpm

       = 1.1 sq ft

- The width is chosen to match the perforated duct width of 2.5'.
  The depth is calculated:

  Minimum non-perforated supply duct depth   = 1.1 sq ft / 2.5'

                                                   = 0.44'

                                                    = 6"

- Non-perforated supply duct cross-sectional dimensions are 2.5' x 6". To avoid errors during construction, it is recommended that the perforated and non-perforated supply ducts have the same depth. The non-perforated supply duct depth is greater and will be used. This reduces the velocity in the perforated duct by increasing the cross sectional area.

Step 7: *Sizing the Intake/Exhaust Area*
- Determine the minimum air intake or exhaust area using Equation 2.9:

- Intake/Exhaust area   = 2,650 cfm / 1,000 fpm

                                = 2.7 sq ft

Chapter 3

Design
Recommendations
and Procedures

Example:
Aeration System for
Cylindrical Bin
(continued)

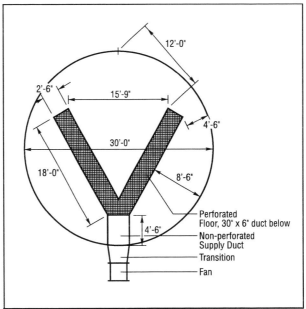

**Figure 3.3. Duct spacing for a Y-shaped duct system in a 30'
diameter bin, filled to 25' level.**

Flat Storage

Flat grain storage is usually inside a building. Even if most of the grain has a level surface, one or more sides might be sloped at the angle of repose. Angle of repose is the angle that results when grain is allowed to flow and come to rest.

The size of aeration ducts in flat storage buildings can be large compared to ducts in round bins because of the larger volume of grain. Design aeration systems so that duct air velocity does not exceed 1,500 fpm and the duct surface air velocity does not exceed 30 fpm. More than one duct at each duct location may be needed to provide adequate duct cross-sectional and/or surface area, or because of the availability and/or cost of different diameter ducts.

Consequently, the specific number of ducts recommended is better thought of as duct locations. For example, two 12-inch diameter ducts might be selected for a duct location, rather than one 18-inch diameter duct or four 12-inch ducts rather than one 24-inch duct, because 12-inch diameter aeration ducts are generally available and less expensive than larger ducts.

Perforated duct length is based on the longer of two criteria:

- Getting air to the ends of the pile without short-circuiting.
- Providing enough perforated surface area for reasonable exit velocity.

For ducts exceeding 100 feet in length, place a fan on both ends of the duct or use a manifold at mid-length, Figure 3.4. For a dual fan system, size each fan and duct for half of the total required airflow. For the center manifold system, size the fan and manifold duct for the total required airflow and the perforated ducts for half the total required airflow. The required duct size is reduced with dual fans. Dual fans could be considered even if a building is less than 100 feet long if equipment cost and/or availability are factors.

A more uniform airflow rate is usually obtained in peaked grain when the ducts are placed parallel to the peak rather than being placed crosswise. A lengthwise duct orientation is usually more economical for narrow building widths.

The information presented in the following sections is based on a maximum sidewall depth of 12 feet and a maximum building width of 100 feet. The airflow path ratio has been maintained at 1.5:1 for most duct locations, Figure 2.9.

The airflow path ratio for a single duct on peaked grain provides adequate airflow because shallow depths and grain near the wall cool naturally. The airflow path ratio for a single duct

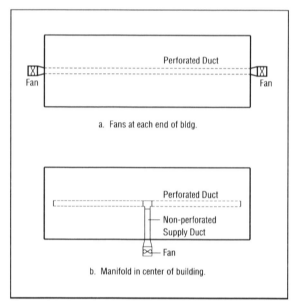

a. Fans at each end of bldg.

b. Manifold in center of building.

Figure 3.4. Aeration for buildings exceeding 100'.
Use fans at each end of the building or a manifold at the center of the building.

system and for the ducts closest to the outside wall can be up to 2:1 for grain depths on the sidewalls up to 4 feet; up to 1.8:1 for greater grain sidewall depths.

Following are design criteria that apply to **all flat storage** aeration systems.

- Positive pressure systems are recommended.
- Maximum perforated duct surface air velocity of 30 fpm.
- Maximum air velocity in perforated duct of 1,500 fpm (1,000 fpm for negative pressure system).
- Maximum air velocity in non-perforated supply duct of 2,500 fpm.
- Maximum intake or exhaust air velocity of 1,000 fpm.
- Maximum duct length of 100 feet. For longer storages, place a fan on each end of the duct or use a manifold in the center of the building.

The following design criteria apply to **flat storage level fill** aeration systems, Figure 3.5:

- Maximum distance between ducts is grain depth.
- Maximum distance between bin wall and duct parallel to wall is one half the grain depth.
- Maximum distance between endwall and end of perforated duct, EP, is 0.7 times the distance between the outside duct and the sidewall.

The following design criteria apply to **flat storage peaked grain** aeration systems:

Lengthwise ducts

For storages up to 60 feet wide, Figure 3.6:
- One duct located at center of storage.
- Maximum distance between endwall and perforated duct, EPP, is 0.7 x (Building width / 2).

For storages 60 to 100 feet wide, Figure 3.7:
- Three ducts.
- Maximum distance between ducts, S3, is Building width / 4.
- Maximum distance between endwall and perforated duct, E3, is 0.7 x S3.

Crosswise ducts, Figure 3.10:
- Maximum distance between ducts, SC, is Building length / (Number of duct locations + 1).
- Maximum distance between sidewall and perforated duct, EPC, is 0.7 x SC.

Apply the following design criteria to **flat storage level top and sloping sides grain** aeration systems, Figure 3.11:

- Maximum distance between duct locations is level grain depth.
- Maximum distance between sidewall and first duct location, SOD, is 0.7 x sloped distance.
- Maximum distance between sidewall and second duct location, S2D, is sloped distance + (level grain depth / 2).

The factor 0.7 for perforated duct spacing from the endwall is determined by making distances B and C, Figure 2.9, equal for an angle of repose of 25 degrees, thus helping to preserve the airflow path ratio design. Published literature (Burrell, 1974; Foster, 1992) uses a factor of 1.0. An analysis of grain volumes served by the end of the duct indicates a factor of 0.5 may be appropriate. The factor 0.7 will be used in the following design examples, because this factor is judged most appropriate.

The design problems that follow use the design criteria presented previously in a step-by-step process. All terminology is defined in the example design problems. A worksheet, Table 3.1, presents these steps and appropriate information to assist with the design process.

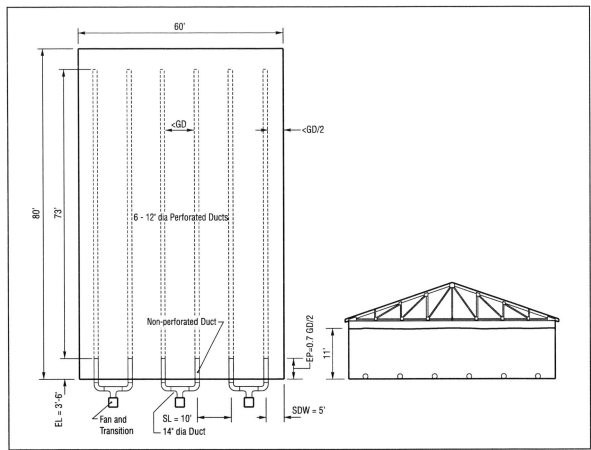

Figure 3.5. Aeration duct spacing and distance from end walls to end of perforated duct.

Table 3.1 Design Guide for Flat Storages

Given Information:
Storage dimensions: _____
Grain type: _____, **Sidewall Grain depth:**___ ft
Maximum Grain depth:___ ft
Design airflow rate: ____ cfm/bu

| | Level Fill | Peaked Filled | | Crosswise Duct | Level Fill, sloping sides |
		0 to 60 feet wide	60 to 100 feet wide		
Step 1					
Number of duct locations	Equation 2.2, pg. 13	1	3	Equation 3.18, pg. 49	Equation 2.2, pg. 13
Duct Spacing, ft	Equation 2.3, pg. 13	centerline	Equation 3.11, pg. 40	Equation 3.17, pg. 48	Equation 2.3, pg. 13
					Equation 3.24, pg. 52
Distance of outside duct from sidewall, ft	Equation 3.1, pg. 36	——————— determined by duct spacing ———————			Equation 3.23, pg. 52
Distance between duct and endwall/sidewall, ft	Equation 3.2, pg. 36	Equation 3.10, pg. 40	Equation 3.12, pg. 40	Equation 3.12, pg. 40	Equation 3.12, pg. 40
Perforated duct length, ft	Equation 3.3, pg. 37	Equation 3.3, pg. 37	Equation 3.3, pg. 37	Equation 3.5, pg. 37	Equation 3.3, pg. 37
Step 2					
Storage Volume, bu	Equation 3.4, pg. 37	Equation 3.13, pg. 44	Equation 3.13, pg. 44	Equation 3.13, pg. 44	Equation 3.25 to Equation 3.32, pg. 53-55
Volume of grain aerated per duct, bu	Equation 3.5, pg. 37	Equation 3.13, pg. 44	Equation 3.14, pg. 44	Equation 3.5, pg. 37	Equation 3.33, pg. 55
			Equation 3.15, pg. 44		Equation 3.34, pg. 56
Step 3					
Airflow per duct, cfm	Equation 3.6, pg. 37	Equation 3.6, pg. 37	Equation 3.6, pg. 37	Equation 3.6, pg. 37	Equation 3.6, pg. 37
Grain Depth, ft	given	Equation 3.16, pg. 45	Equation 3.16, pg. 45	Equation 3.19, pg. 50	Equation 3.35, pg. 56
					Equation 3.16, pg. 45

Static Pressure   Table 2.1, pg. 7; Use fan airflow for the following steps.

Step 4
Choosing a Fan from Manufacturer's Data (Similar to Table 2.2)

Step 5

Duct size, sq ft or diameter   Equation 3.7, pg. 38

Step 6
Minimum perforated duct length, ft   Equation 3.8, pg. 38

Step 7
Total required airflow, cfm   Equation 3.9, pg. 39
Size exhaust are, sq ft   Equation 2.9, pg. 21

Chapter 3

Design
Recommendations
and Procedures

Example:
Level-filled flat storage

Example 3.3: Level-filled flat storage

Given Information:
Storage dimensions: <u>60 x 80</u> ft
Grain type: <u>Shelled Corn</u>, **Grain depth:** <u>11</u> ft
Design airflow rate: <u>0.10</u> cfm/bu

SOLUTION:
Because the building is 80' long, it makes sense to use lengthwise ducts, Figure 3.5. This is temporary storage so above-floor round ducts will be used.

Step 1: *Calculating Duct Spacing*
• Using Equation 2.2, remember to round up:
• Number of duct locations   = building width/grain depth
                                                    = 60' / 11'
                                                    = 5.5
                                                    = 6 duct locations

Equation 2.3 Duct spacing, level fill

$$SL = \frac{W \text{ or } BD}{ND}$$

Where: S   = Duct spacing, ft
              ND = Number of ducts
              BD = Bin diameter, ft
              W   = Building width, ft

SL = 60' / 6 ducts
    = 10'

Equation 3.1: Distance from sidewall to duct

$$SDW = \frac{SL}{2}$$

Where: SDW = Distance from sidewall to first duct, ft
              SL   = Duct spacing, ft

SDW = 10' / 2
       = 5'

Equation 3.2: Endwall distance to perforated duct

EP = 0.7(SDW)

Where: EP = Distance from endwall to perforated duct, ft
             SDW = Distance from sidewall to first duct, ft

EP =   0.7(5)
    =   3.5'
    =   3'-6"

Chapter 3

Design
Recommendations
and Procedures

Example:
Level-filled flat storage
(continued)

Equation 3.3: Perforated Duct Length

PDL = L-2(EP)

Where: PDL = Perforated duct length, ft
       L   = Building length, ft
       EP = Distance from endwall to perforated duct, ft

PDL = 80' - (2(3.5'))
      73'

Step 2: *Calculating the Storage Volume*
• Estimate the amount of grain in the storage, and the amount of grain served by each duct location.

Equation 3.4: Storage Volume

TV = (VG)(GD)

Where: TV   = Total storage volume, bu
        VG = Volume of grain per foot of depth, bu/ft (Table 3.2)

TV = 3,840 bu/ft x 11'
    = 42,240 bu

Equation 3.5: Volume of grain aerated by each duct location

$$GAD = \frac{TV}{ND}$$

Where: GAD = Amount of grain aerated per duct, bu
        TV   = Total storage volume, bu
        ND = Number of ducts

$$GAD = \frac{42,240 \text{ bu}}{6 \text{ ducts}}$$

      = 7,040 bu/duct

Step 3: *Calculating Airflow*

Equation 3.6: Airflow per duct location

APD = (AR)(GAD)

Where: APD = Airflow per duct, cfm
        GAD = Amount of grain aerated per duct, bu
        AR = Airflow rate, cfm/bu

APD = 0.10 cfm/bu x 7,040 bu/duct
     = 704 cfm/duct

Design
Recommendations
and Procedures

Example:
Level-filled flat storage
(continued)

- Estimate the depth of grain over each duct and the static pressure based on the design airflow. The estimated static pressure is determined using Table 2.1:

  Grain depth = 11.0'
  Static pressure = 0.6"

  Use the 15 ft value rather than the 10 ft value to be conservative and since there is only a minor difference in static pressure values.

Step 4: *Choosing a Fan*
- Choose a fan by comparing fan manufacturers' data like Table 2.2.
  Axial, 12" dia., 0.5 hp, 1,500 cfm at a static pressure of 1" for every two ducts using a manifold.

Step 5: *Choosing Ducts*
- Select the duct sizes from the information presented in Chapter 2, **System Components**. If the perforated duct will be round or semicircular, select the duct size using Tables 2.5 and 2.6. If the aeration fan selected moves more air than the airflow the calculations indicate, select ducts that will accommodate the actual airflow the fan provides. Maximum allowable duct air velocity is 1,500 fpm. For the example, round ducts will be sized as follows:
  Airflow/duct location = 1,500 cfm / 2 = 750 cfm/duct

Equation 3.7: Duct Area

$$DA = \frac{APD}{1,500}$$

Where: DA = Recommended duct area, sq ft
APD = Airflow per duct, cfm

DA = 750 cfm / 1,500 fpm
= 0.5 sq ft
= 10" round duct
- Because 10" round duct is not commonly available, use a 12" diameter duct.

Supply duct airflow = 1,500 cfm
Duct area = 1,500 cfm / 2,500 fpm
= 0.6 sq ft
- Use a 12" round duct (0.79 sq ft)

Step 6: *Determining Perforated Duct Length*
- Determine the minimum perforated duct length by checking the duct surface air velocity for the perforated section of the ducts. The duct surface area is found using Table 2.5. Maximum allowable perforated duct surface air velocity is 30 fpm.

- Surface area = 2.51 sq ft/ft (length), from Table 2.5

Equation 3.8: Perforated Duct Length

$$PDL_{min} = \frac{APD}{(SA)(30)}$$

Where: PDL_{min} = Minimum perforated duct length, ft
SA = Surface area, sq ft/ft length
APD = Airflow per duct, cfm

Chapter 3

Design
Recommendations
and Procedures

Example:
Level-filled flat storage
(continued)

$$PDL_{min} = 750 \text{ cfm} / (2.51 \text{ sq ft/ft} \times 30 \text{ fpm})$$
$$= 10'$$

• Note that the length for perforated sections, 73', calculated in Step 1 is sufficient to meet this design criteria.

Step 7: *Calculating Total Airflow*

Equation 3.9: **Determine the total airflow**

$$Q_{req'd} = (APD)(ND)$$

Where: $Q_{req'd}$ = Total airflow, cfm
APD = Airflow per duct, cfm
ND = Number of ducts

$$Q_{req'd} = 750 \text{ cfm/duct} \times 6 \text{ ducts}$$
$$= 4{,}500 \text{ cfm}$$

• Determine minimum air exhaust area needed for the grain storage building by using Equation 2.9:

Exhaust area = 4,500 cfm / 1,000 fpm
= 4.5 sq ft

Table 3.2. Estimated grain volume per foot of grain depth in a flat storage with a level top.

GV = LW 0.8

Grain Storage length, ft	Storage width, ft										
	20	28	36	44	52	60	68	76	84	92	100
	bu/ft of grain depth										
20	320										
24	384										
28	448	627									
32	512	717									
36	576	806	1,037								
40	640	896	1,152								
44	704	986	1,267	1,549							
48	768	1,075	1,382	1,690							
52	832	1,165	1,498	1,830	2,163						
56	896	1,254	1,613	1,971	2,330						
60	960	1,344	1,728	2,112	2,496	2,880					
64	1,024	1,434	1,843	2,253	2,662	3,072					
68	1,088	1,523	1,958	2,394	2,829	3,264	3,699				
72	1,152	1,613	2,074	2,534	2,995	3,456	3,917				
76	1,216	1,702	2,189	2,675	3,162	3,648	4,134	4,621			
80	1,280	1,792	2,304	2,816	3,328	3,840	4,352	4,864			
84	1,344	1,882	2,419	2,957	3,494	4,032	4,570	5,107	5,645		
88	1,408	1,971	2,534	3,098	3,661	4,224	4,787	5,350	5,914		
92	1,472	2,061	2,650	3,238	3,827	4,416	5,005	5,594	6,182	6,771	
96	1,536	2,150	2,765	3,379	3,994	4,608	5,222	5,837	6,451	7,066	
100	1,600	2,240	2,880	3,520	4,160	4,800	5,440	6,080	6,720	7,360	8,000

Chapter 3

Design
Recommendations
and Procedures

Flat Storage,
Peaked Grain:
Lengthwise Ducts

Flat Storage, Peaked Grain: Lengthwise Ducts

This section provides separate design procedures for storages up to 60 feet wide and for storages from 60 to 100 feet wide.

Storages up to 60 feet wide

One properly sized duct located lengthwise under the grain peak is generally sufficient for grain piled **without flat surfaces**, for buildings up to 60 feet wide, Figure 3.6.

Grain depth on the sidewall up to 12 feet will not affect the duct spacing. Use the following formula to determine the distance from the endwall to the end of the perforated duct:

Equation 3.10: Distance from endwall to perforated duct

$$EPP = 0.7 \frac{W}{2}$$

Where: EPP = Distance from endwall to perforated duct, peaked grain, ft

  W = Building width, ft

Storages 60 to 100 feet wide

For buildings 60 to 100 feet wide, use three ducts running lengthwise. One duct is located in the center of the building. The other two are located one-fourth of the building width from each sidewall, Figure 3.7. Grain depth on the sidewall up to 12 feet will not affect the duct spacing. Use the following formulas to determine spacing between duct locations and the distance from the endwall to the end of the perforated duct:

Equation 3.11: Duct spacing, three ducts

$$S3 = \frac{W}{4}$$

Where: S3 = Spacing between ducts, ft
  W = Building width, ft

Equation 3.12: Distance from endwall to perforated duct, three ducts

$$E3 = 0.7(S3)$$

Where: E3 = Distance from endwall to perforated duct, ft
  S3 = Spacing between ducts, ft

When the grain storage building is nearly square, the spacing of the perforated section of the duct from the endwall calculated using Equation 3.10 or 3.12 may result in a duct that is too short to satisfy the outlet area criteria. In these cases, use the minimum length of perforated duct as calculated in Step 6.

Chapter 3

Design
Recommendations
and Procedures

Flat Storage,
Peaked Grain:
Lengthwise Ducts

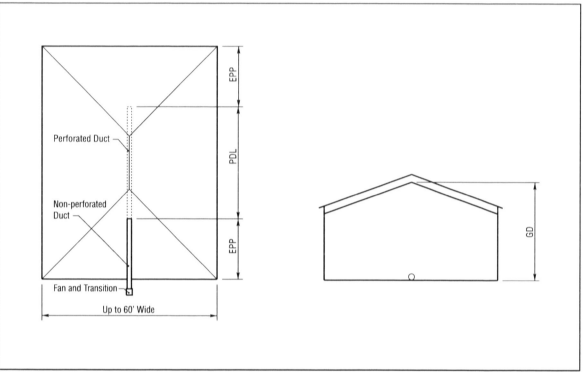

Figure 3.6. Up to 60-foot wide flat storage requires one duct.
GD = Grain depth

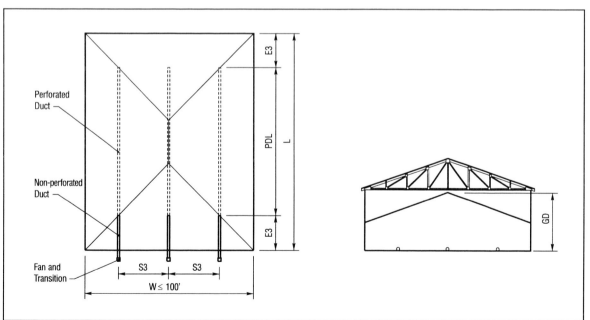

Figure 3.7. Up to 100-foot wide flat storage requires three ducts.
GD = Grain depth

Design
Recommendations
and Procedures

Flat Storage,
Peaked Grain:
Lengthwise Ducts

Table 3.3. Estimated grain volume in a flat storage with a peaked top with no grain on the sidewalls, bu.

Angle of repose = 25 degrees

Grain storage length, ft	Grain storage width, ft										
	20	28	36	44	52	60	68	76	84	92	100
20	501										
24	652										
28	802	1,376									
32	953	1,670									
36	1,103	1,965	2,924								
40	1,253	2,260	3,411								
44	1,404	2,555	3,898	5,338							
48	1,554	2,850	4,386	6,066							
52	1,705	3,144	4,873	6,794	8,811						
56	1,855	3,439	5,360	7,522	9,828						
60	2,005	3,734	5,848	8,250	10,845	13,536					
64	2,156	4,029	6,335	8,978	11,862	14,890					
68	2,306	4,323	6,822	9,706	12,878	16,243	19,704				
72	2,457	4,618	7,309	10,434	13,895	17,597	21,443				
76	2,607	4,913	7,797	11,162	14,912	18,950	23,182	27,509			
80	2,757	5,208	8,284	11,890	15,928	20,304	24,920	29,681			
84	2,908	5,503	8,771	12,618	16,945	21,658	26,659	31,853	37,143		
88	3,058	5,797	9,259	13,345	17,962	23,011	28,398	34,024	39,796		
92	3,209	6,092	9,746	14,073	18,978	24,365	30,136	36,196	42,449	48,798	
96	3,359	6,387	10,233	14,801	19,995	25,718	31,875	38,368	45,102	51,980	
100	3,509	6,682	10,721	15,529	21,012	27,072	33,613	40,540	47,755	55,163	62,667

Value in box is used in Example 3.4, Flat Storage, Peaked Grain, Step 2.

Table 3.4. Estimated portion of the grain volume in a flat storage with a peaked top served by each of the outside ducts in a three duct aeration system, bu.

Angle of repose = 25 degrees.

Storage sidewall depth, ft	Storage width, ft										
	20	28	36	44	52	60	68	76	84	92	100
1	0.10	0.09	0.08	0.08	0.07	0.07	0.07	0.07	0.07	0.07	0.07
2	0.14	0.12	0.11	0.10	0.10	0.09	0.09	0.09	0.09	0.09	0.09
3	0.16	0.14	0.13	0.12	0.12	0.11	0.11	0.11	0.10	0.10	0.10
4	0.17	0.16	0.15	0.14	0.13	0.13	0.12	0.12	0.12	0.11	0.11
5	0.18	0.17	0.16	0.15	0.14	0.14	0.13	0.13	0.13	0.13	0.12
6	0.19	0.18	0.17	0.16	0.15	0.15	0.14	0.14	0.14	0.13	0.13
7	0.20	0.19	0.18	0.17	0.16	0.16	0.15	0.15	0.14	0.14	0.14
8	0.20	0.19	0.18	0.17	0.17	0.16	0.16	0.15	0.15	0.15	0.15
9	0.21	0.20	0.19	0.18	0.17	0.17	0.16	0.16	0.16	0.16	0.15
10	0.21	0.20	0.19	0.18	0.18	0.17	0.17	0.17	0.16	0.16	0.16
11	0.21	0.20	0.20	0.19	0.18	0.18	0.17	0.17	0.17	0.17	0.16
12	0.22	0.21	0.20	0.19	0.19	0.18	0.18	0.18	0.17	0.17	0.17

Value in box is used in Example 3.4, Flat Storage, Peaked Grain, Step 2.

Chapter 3

Design
Recommendations
and Procedures

Example:
Lengthwise duct
system for peaked
storage

Example 3.4: Lengthwise duct system for peaked storage

Given Information:
Storage dimensions: 60' x 80'
Grain type: __Shelled Corn__ , **Sidewall grain depth:** 4 ft
Design airflow rate: 0.10 cfm/bu

SOLUTION:

Step 1: *Choosing the number of ducts*
- Select the number of duct locations based on the size and geometry of the grain storage. For this example, either a single duct system, Figure 3.8, or a three duct system, Figure 3.9, can be used.

- Both duct systems have a duct placed in the center of the building. Duct spacing for the three duct system is calculated using Equation 3.11 as:
$$S3 = 60' / 4$$
$$= 15'$$

- Determine the distance from the end wall to the perforated section and the length of the perforated duct:
Single duct using Equation 3.10:
$$EPP = 0.7 \times (60' / 2)$$
$$= 21'$$

- Use 20' non-perforated supply duct because that is a common length.
Using Equation 3.3:
$$PDL = 80' - (2(20'))$$
$$= 40'$$

- Three duct design using Equation 3.12:
$$E3 = 0.7 \times 15'$$
$$= 10.5'$$
$$= 11'$$

- Using Equation 3.3:
$$PDL = 80' - (2(11'))$$
$$= 58'$$

- Use 60' length because it is a commonly available duct length.

Step 2: *Calculating Storage Volume*
- Estimate the amount of grain in the storage and the amount of grain served by each duct location.

- The amount of grain in the 4' section at the bottom of the pile, Figure 3.9, can be estimated using Table 3.2, which gives 3,840 bu/ft of grain depth. The amount of grain in the peaked section can be estimated using Table 3.3, which gives 20,304 bu. Therefore, the total estimated volume of grain in the storage is:

Chapter 3

Design
Recommendations
and Procedures

Flat Storage,
Peaked Grain:
Lengthwise Ducts
(continued)

Equation 3.13: Storage Volume

$$SV = (BLV)(SWD) + PV$$

Where:
$$
\begin{aligned}
SV &= \text{Storage Volume, bu} \\
BLV &= \text{Bottom Level Volume, bu (Table 3.2)} \\
SWD &= \text{Sidewall Depth, ft} \\
PV &= \text{Peaked Volume, bu (Table 3.3)}
\end{aligned}
$$

$$
\begin{aligned}
SV &= (3{,}840 \text{ bu/ft} \times 4') + 20{,}304 \text{ bu} \\
&= 35{,}664 \text{ bu}
\end{aligned}
$$

- Estimate the amount of grain served by each duct. For a single duct system, the center duct must serve all 35,664 bu. For a three duct system, the percentage of the total grain served by each of the outside duct locations can be estimated using the value from Table 3.4 in Equation 3.14 below:

Equation 3.14: Grain aerated by outside duct

$$VGOD = C1(SV)$$

Where:
$$
\begin{aligned}
VGOD &= \text{Volume of grain aerated by outside duct, bu/duct} \\
C1 &= \text{Coefficient from Table 3.4} \\
SV &= \text{Storage volume, bu}
\end{aligned}
$$

$$
\begin{aligned}
VGOD &= 0.13 \times 35{,}664 \text{ bu} \\
&= 4{,}636 \text{ bu/duct}
\end{aligned}
$$

Equation 3.15: Grain aerated by center duct

$$VGCD = SV - 2(VGOD)$$

Where:
$$
\begin{aligned}
VGCD &= \text{Volume of grain aerated by center duct, bu/duct} \\
VGOD &= \text{Volume of grain aerated by outside duct, bu/duct} \\
SV &= \text{Storage volume, bu}
\end{aligned}
$$

$$
\begin{aligned}
VGCD &= 35{,}664 \text{ bu} - (2 \text{ ducts} \times 4{,}636 \text{ bu/duct}) \\
&= 26{,}392 \text{ bu/duct}
\end{aligned}
$$

Step 3: *Calculating Airflow*
- Calculate the airflow for each duct location using

Equation 3.6: Airflow

Single duct:
$$
\begin{aligned}
\text{Airflow/duct location} &= 0.10 \text{ cfm/bu} \times 35{,}664 \text{ bu} \\
&= 3{,}566 \text{ cfm/duct}
\end{aligned}
$$

Three duct:
Outside:
$$
\begin{aligned}
\text{Airflow/duct location} &= 0.10 \text{ cfm/bu} \times 4{,}636 \text{ bu} \\
&= 464 \text{ cfm/duct}
\end{aligned}
$$

Center:
$$
\begin{aligned}
\text{Airflow/duct location} &= 0.10 \text{ cfm/bu} \times 26{,}392 \text{ bu} \\
&= 2{,}639 \text{ cfm/duct}
\end{aligned}
$$

Chapter 3

Design
Recommendations
and Procedures

Flat Storage,
Peaked Grain:
Lengthwise Ducts
(continued)

- Estimate the depth of grain over each duct and the static pressure based on the design airflow. The depth of grain above the sidewall height over ducts in a flat storage is provided in Table 3.5. Grain depth is calculated with Equation 3.16. The estimated static pressure is determined using Table 2.1:

Equation 3.16: Grain depth over duct

$$GD = GDP + SWD$$

Where: GD   = Grain depth, ft
           GDP = Peaked grain depth over duct, ft (Table 3.5)
           SWD = Grain depth at sidewall, ft

    GD = 14' + 4.0'
        = 18'

- Static pressure   = 0.7"

Three duct:
- Center grain depth = 14' + 4.0'
                          = 18'

   Static pressure   = 0.7"

- Outer grain depth   = 7.0' + 4.0'
                          = 11.0'

   Static pressure   = 0.5"

Step 4: *Choosing a Fan*
- Choose a fan by comparing fan manufacturers' data like that in Table 2.2.

Single: Axial, 16" dia.; 1.5 hp; 3,500 cfm at a static pressure of 1"

Three: *Outer . . . Axial; 12" dia.; 0.5 hp, 1,500 cfm at a static pressure of 1"
        Center . . Axial; 14" dia.; 1.0 hp, 2,880 cfm at a static pressure of 1"
*NOTE: This is the smallest fan in the table and therefore will be used in this
        example. Fans more closely matching design airflow exist and should be used.

Step 5: *Choosing Ducts*
- Select the duct sizes from the information presented in Chapter 2, **System Components**. If the perforated duct will be round or semicircular, select the duct size using Tables 2.5 and 2.6. If the aeration fan selected moves more air than the airflow used in the calculations, select ducts that accommodate the actual airflow provided by the fan. Maximum allowable duct air velocity is 1,500 fpm. For this example, round ducts will be sized as follows using Equation 3.7:

Single duct:
    Airflow per duct location = 3,500 cfm

    Duct area   = 3,500 cfm / 1,500 fpm
                                   = 2.33 sq ft
                                   = 21" round duct

Chapter 3

Design
Recommendations
and Procedures

Flat Storage,
Peaked Grain:
Lengthwise Ducts
(continued)

Three duct:
Outside duct area = 1,500 cfm / 1,500 fpm
= 1.0 sq ft
= 14" round duct

Because the fan selected provides more airflow than needed, choose a 12" diameter duct to match fan size. Realize that the increased velocity will increase the static pressure and decrease the delivered airflow.

Center duct area = 2,880 cfm / 1,500 fpm
= 1.92 sq ft
= 18" round duct (chosen because it is the common duct size with a cross-sectional area closest to the calculated size)

Step 6: *Calculating Duct Length*
- Determine the minimum perforated duct length by checking the duct surface air velocity for the perforated section of the ducts. The duct surface area is found using Table 2.5. Maximum allowable perforated duct surface air velocity is 30 fpm. Using Equation 3.8:

12" duct:
Surface area = 2.51 sq ft/ft, from Table 2.5
Minimum length = 1,500 cfm/(2.51 sq ft/ft x 30 fpm)
= 19.9'
18" duct:
Surface area = 3.77 sq ft/ft
Minimum length = 2,880 cfm / (3.77 sq ft/ft x 30 fpm)
= 25.5'
21" duct:
Surface area = 4.40 sq ft/ft
Minimum length = 3,500 cfm / (4.40 sq ft/ft x 30 fpm)
= 26.5'
- Note that the length calculated in Step 1, 58' is sufficient to meet the design criteria for the single-duct and three-duct plan.

Step 7: *Sizing Exhaust Area*
- Determine the total airflow and minimum air exhaust area needed for the grain storage building using Equations 3.9 and 2.9:

Single duct:
Total airflow = 3,500 cfm, from Step 3.

Exhaust area = 3,500 cfm / 1,000 fpm
= 3.5 sq ft
Three duct:
Total airflow = (2 x 1,500 cfm) + 2,880 cfm
= 5,880 cfm

Exhaust area = 5,880 cfm / 1,000 fpm
= 5.9 sq ft

Chapter 3

Design
Recommendations
and Procedures

Flat Storage,
Peaked Grain:
Lengthwise Ducts
(continued)

Table 3.5. Depth of grain above the sidewall height over ducts for flat grain storage with peaked top, ft.

Angle of repose = 25 degrees $D_c = 0.233\,W$ $D_o = D_c / 2$

Duct location	Grain storage width, ft										
	20	28	36	44	52	60	68	76	84	92	100
Center	5	7	8	10	12	14	16	18	20	21	23
Outside	2	3	4	5	6	7	8	9	10	11	12

Values highlighted in the boxes used in Example 3.4, Flat Storage, Peaked Grain, Step 3.

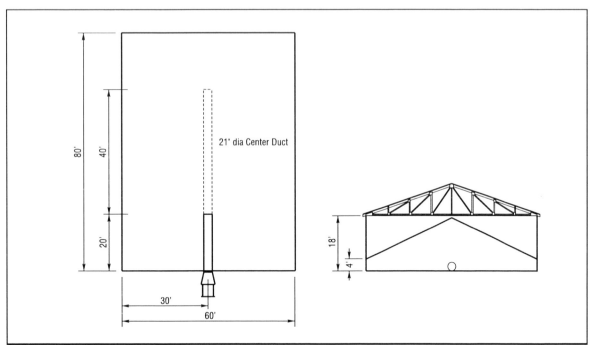

Figure 3.8. Flat storage with peaked grain-single duct.

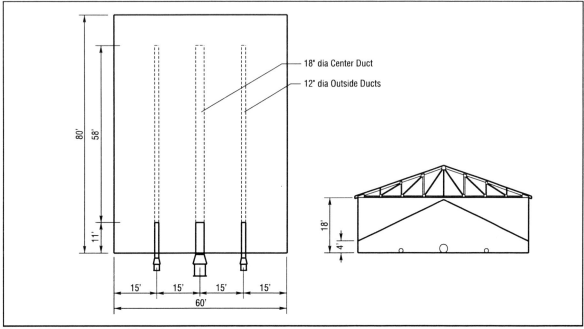

Figure 3.9. Flat storage with peaked grain, three ducts

Chapter 3

Design
Recommendations
and Procedures

Example:
Crosswise duct system
for Peaked Storage

Example 3.5: Crosswise duct system for peaked storage

Given Information:
Storage dimensions: <u>60 ft x 100 ft</u>
Grain type: <u>Shelled Corn</u> , **Sidewall grain depth:** <u>4</u> ft
Design airflow rate: <u>0.10</u> cfm/bu

SOLUTION:

Step 1: *Calculating Duct Spacing*
- When running ducts crosswise in a flat storage, Figure 3.10, the number of ducts and the spacing between duct locations is determined from the sidewall depth as given in Equation 3.17. The number of duct locations is calculated using Equation 3.18. Because the grain is sloping at both ends of the pile, one less duct is used than would be used for a level pile depth. Always increase the number of duct locations for a calculated fraction of duct locations.

Equation 3.17: Duct Spacing

$$SC = 2 + 2(SWD)$$

Where: SC = Duct spacing for crosswise duct layout, ft
         SWD = Grain depth at sidewall, ft

$$SC = 2 + (2(4)) = 10 \text{ ft}$$

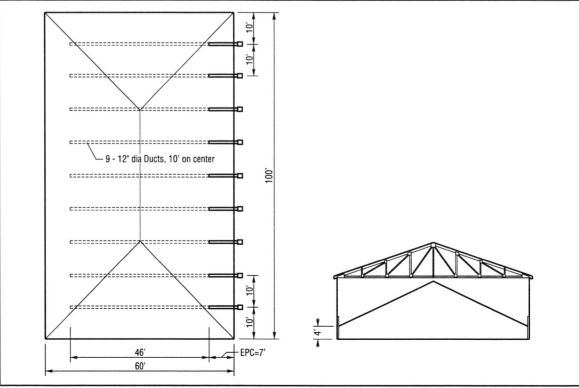

Figure 3.10. Ducts crosswise in a flat storage with peaked grain.

Chapter 3

Design
Recommendations
and Procedures

Example:
Crosswise duct system
for Peaked Storage
(continued)

Equation 3.18: Number of Ducts

$$ND = \frac{L}{SC} \; 1$$

Where: ND = Number of ducts
SC = Duct spacing for crosswise duct layout, ft
L = Building length, ft

$$ND = \frac{100'}{10'} - 1$$

$$= 9$$

• Determine the distance from the sidewall to the perforated section and the length of the perforated duct. The distance from the sidewall and the length of the perforated section of each duct are calculated using Equations 3.12 and 3.3 with building width substituted for building length:

EPC = 0.7 x SC
= 7'
EPC = Distance from sidewall to perforated duct, crosswise ducts, ft.

PDL = 60' - (2 x7')
= 46'

Step 2: *Calculating Storage Volume*
• Estimate the amount of grain in the storage and the amount of grain served by each duct location. The amount of grain in the 4' section at the bottom can be estimated using Table 3.2, i.e. 4,800 bu/ft of grain depth. Estimate the amount of grain in the peaked section using Table 3.3, i.e. 27,072 bu. Therefore, the total estimated volume of grain using Equation 3.13 is:

• Storage Volume = (4,800 bu/ft x 4') + 27,072 bu
= 46,272 bu

• Volume of grain aerated by each duct location using Equation 3.5:

• Volume per duct = 46,272 bu / 9 duct
= 5,141 bu per duct

• The ducts nearest the endwall serve slightly less grain, but the design is still acceptable.

Step 3: *Calculating Airflow*

Equation 3.6: Airflow for each duct
Airflow/duct = 0.10 cfm/bu x 5,141 bu
= 514 cfm/duct

• Estimate the depth of grain over each duct and the static pressure based on the desired airflow. The depth of grain above the sidewall height over ducts in a flat grain storage is provided in Table 3.5. The average grain depth between peak and sidewall is used to determine the static pressure. Note that a variation in airflow will exist as the grain depth varies along the length of the duct.

Chapter 3

Design
Recommendations
and Procedures

Example:
Crosswise duct system
for Peaked Storage
(continued)

Equation 3.19: Average Grain Depth

$$GD_{avg} = \frac{GDP}{2} + SWD$$

Where: GD_{avg} = Average grain depth, ft
GDP = Peak grain depth, ft (Table 3.5)
SWD = Grain depth at sidewall, ft

$$GD_{avg} = (14/2) + 4$$
$$= 11'$$

• The estimated static pressure is determined using Table 2.1:

Static pressure = 0.6"

Step 4: *Choosing a Fan*
• Choose a fan by comparing fan manufacturers' data like Table 2.2.

• Axial 0.5 hp; 12" dia.; 1,500 cfm at static pressure of 1"

NOTE: This is the smallest fan in the table. Fans more closely matching design airflow exist. Also consider using a manifold so one fan serves more than one duct.

Step 5: *Choosing Ducts*
• Select the duct size from the information presented in Chapter 2, **System Components**. If the perforated duct will be round or semicircular, select the duct size using Tables 2.5 and 2.6. If the aeration fan selected moves more air than the airflow used in the calculations, select ducts that will accommodate the actual airflow provided by the fan. Maximum allowable duct air velocity is 1,500 fpm. For this example, round ducts will be sized using Equation 3.7:
Airflow/duct   = 1,500 cfm/duct

Duct area   = 1,500 cfm/duct / 1,500 cfm/sq ft
= 1.0 sq ft
= 14" round duct

Step 6: *Duct Length*
• Determine the minimum perforated duct length by checking the duct surface air velocity for the perforated section of the ducts. The duct surface area is found using Table 2.5.
Minimum length = 1,500 / 88 =17'

• Note that the length for perforated sections as calculated in Step 1 is sufficient to meet this design criteria.

Step 7: *Calculating Exhaust Area*
• Determine the total airflow and minimum air exhaust area needed for the grain storage building using Equations 3.9 and 2.9:

Total airflow   = 1,500 cfm/duct x 9 ducts
= 13,500 cfm

Exhaust area   = 13,500 cfm / 1,000 fpm
= 13.5 sq ft

Chapter 3

Design
Recommendations
and Procedures

Flat Storage with level
top and sloping sides.

Flat Storage with Level Top and Sloping Sides

Part of the grain surface in the middle of flat storage buildings may be level, and the remainder of the surface sloped to the grain walls or bulkheads, Figure 3.11. This situation is most common in storage structures with truss supported roofs. Do not allow stored grain to touch the lower chord of trusses. If grain moves against the lower chord during loading, unloading, or natural grain shifting, the lower chords could become damaged or broken, which could result in roof damage or collapse.

Having both level and sloped grain surfaces in a building is not undesirable. An important advantage is that the level grain surfaces in the middle of the building are easy to walk on when inspecting the grain's condition.

The design of a perforated duct positive pressure aeration system for a storage having both level and peaked grain surfaces uses the procedure for the level surfaces for that portion of grain and the procedure for peaked surfaces for the sloped grain.

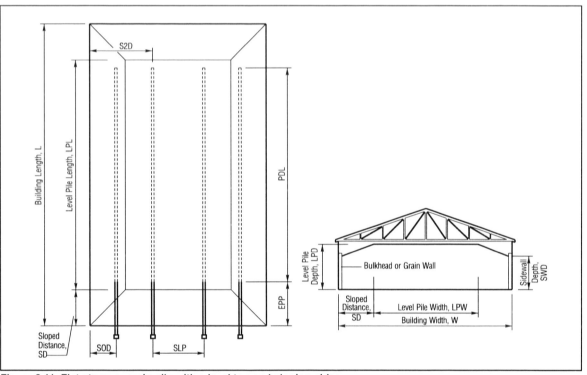

Figure 3.11. Flat storage: grain pile with a level top and sloping sides.

Example 3.6: *Flat storage with level top and sloping sides*

Given Information:
Storage dimensions: Bin dia <u>80' x 100'</u>
Grain type: <u>Shelled Corn</u>, **Sidewall grain depth:** <u>4</u> ft
LPD: <u>16</u> ft
Design airflow rate: <u>0.10</u> cfm/bu

SOLUTION:

Step 1: *Calculating the Number of Ducts, Spacing, and Perforated length*
- Determine the number of duct locations, the spacing between the ducts, the spacing of the outside ducts from the sidewall, the spacing of the perforated section of the duct from the end wall of the storage, and the length of the perforated section of the duct. Grain pile dimensions are calculated first by the following equations:

Design
Recommendations
and Procedures

Example:
Flat Storage with level
top and sloping sides.
(continued)

Equation 3.20: Length of Sloped Section

$$SD = 2.14(LPD - SWD)$$

Where: SD = Horizontal length of the sloped section of grain, ft
LPD = Level pile depth, ft
SWD = Grain depth at sidewall, ft

• Calculate the width and length of the level surface grain pile:

Equation 3.21: Level Pile Width

$$LPW = W - 2(SD)$$

Where: LPW = Level pile width, ft
SD = Horizontal length of the sloped section of grain, ft
W = Building width, ft

Equation 3.22: Level Pile Length

$$LPL = L - 2(SD)$$

Where: LPL = Level pile length, ft
L = Building length, ft
SD = Horizontal length of the sloped section of grain, ft

Equation 3.23: Distance of outside duct location from sidewall

$$SOD = 0.7(SD)$$

Where: SOD = Distance from outside duct to sidewall, ft
SD = Horizontal length of the sloped section of grain, ft

Equation 3.24: Distance of second duct location from sidewall

$$S2D = SD + \frac{SLP}{2}$$

Where: S2D = Distance from second duct to sidewall, ft
SLP = Duct spacing for the level section of grain, ft
SD = Horizontal length of the sloped section of grain, ft

• Estimate the horizontal length of the sloped grain (the distance from the edge of the level grain to the grain wall or bulkhead), level surface width and length using Equations 3.20 to 3.22:

$$
\begin{aligned}
SD \quad &= \quad 2.14 \times (16'\text{-}4') \\
&= \quad 25.7' \\[6pt]
LPW \quad &= \quad 80' - (2 \times 25.7') \\
&= \quad 28.6' \\
&= \quad 28'\text{-}7'' \\[6pt]
LPL \quad &= \quad 100' - (2 \times 25.7') \\
&= \quad 48.6' \\
&= \quad 48'\text{-}7''
\end{aligned}
$$

Chapter 3

Design
Recommendations
and Procedures

Example:
Flat Storage with level
top and sloping sides.
(continued)

- Select the number of duct locations required and the location. The maximum distance between duct locations for the level portion of the grain is equal to the grain depth. Always increase the number of duct locations for calculated fraction of ducts.

- Using Equation 2.2 with the terms from Figure 3.11:

$$
\begin{aligned}
\text{Number of duct locations} &= \text{LPW / LPD} \\
&= 28.6' / 16' \\
&= 1.8 \\
&= 2 \text{ duct locations}
\end{aligned}
$$

- Calculate the duct location spacing using Equation 2.3 with the terms from Figure 3.11:

$$
\begin{aligned}
\text{SLP} &= \text{LPW / Number of ducts in level top section} \\
&= 28.6' / 2 \\
&= 14.3' \\
&= 14'
\end{aligned}
$$

- Calculate the distance of the outside duct from each sidewall in the sloping portion of the grain using Equation 3.23:

$$
\begin{aligned}
\text{SOD} &= 0.7(25.7') \\
&= 18'
\end{aligned}
$$

- Calculate the distance of the second duct location from the sidewall using Equation 3.24:

$$
\begin{aligned}
\text{S2D} &= 25.7' + (14' / 2) \\
&= 32.7' \\
&= 33'
\end{aligned}
$$

- Calculate the distance from the endwall and length of the perforated section of the ducts using Equations 3.12 and 3.3:

$$
\begin{aligned}
\text{ESL} &= 0.7(14) \text{ (based on center duct spacing)} \\
&= 9.8' \\
&= 10' \\
\text{PDL} &= 100' - (2(10')) \\
&= 80'
\end{aligned}
$$

Step 2: *Calculating Storage Volume*

- Estimate the amount of grain in the storage and the amount of grain served by each duct location. The volume can be estimated by breaking the storage into three separate volumes, Figure 3.12. Using the formula given in Table 3.2, the volume of the level top center section is calculated. The dimensions from Figure 3.13 are used as follows:

Equation 3.25: Center Section Grain Volume

$$\text{CSV} = 0.8(\text{LPW})(\text{LPL})(\text{LPD})$$

Where:
CSV	=	Volume of grain in the center section, bu
LPW	=	Level pile width, ft
LPL	=	Level pile length, ft
LPD	=	Level pile depth, ft

Design
Recommendations
and Procedures

Example:
Flat Storage with level
top and sloping sides.
(continued)

- The amount of grain in the sidewall section, Figure 3.12, needs to be calculated using a two step process. The first step determines the amount of grain in the sidewall slopes, (C) and (E). The second step determines the amount of grain in the sidewall sections from ground level to the sidewall grain height, (B) and (D).

Equation 3.26: Volume of Grain in Sidewall Sloped Section

$$SSV = 0.8(0.47)\left[\left(SD^2\right)(LPL) + \left(\frac{4\left(SD^3\right)}{3}\right)\right]$$

Where: SSV = Volume of grain in the sloped side sections, bu
SD = Horizontal length of the sloped section of grain, ft
LPL = Level pile length, ft

Equation 3.27: Sidewall Depth Section Grain Volume

$$SDV = (2)(0.8)(L)(SD)(SWD)$$

Where: SDV = Volume of grain in the sidewall height portion of the sloped section, bu
SD = Horizontal length of the sloped section of grain, ft
SWD = Grain depth at sidewall, ft
L = Building length, ft

Equation 3.28: Total Volume of Grain in the Sidewall Sections

$$TSV = SSV + SDV$$

Where: TSV = Total volume of grain in the side section, bu
SSV = Volume of grain in the sloped side sections, bu
SDV = Volume of grain in the sidewall height portion of the sloped section, bu

- Similarly, the two step method is used to estimate the amount of grain in the endwall section, Figure 3.12. The first step determines the amount of grain in the endwall slopes, (G) and (I). The second part determines the amount of grain in the wall sections from ground level to the sidewall grain height, (F) and (H).

Equation 3.29: Volume of Grain in Sloped Section by Endwall

$$ESV = 0.8(0.47)(SD^2)(LPW)$$

Where: ESV = Volume of grain in the sloped sections next to the endwalls, bu
SD = Horizontal length of the sloped sections of grain, ft
LPW = Level pile width, ft

Equation 3.30: Volume of Grain in the Endwall Depth Section

$$EDV = 2(0.8)(SD)(LPW)(SWD)$$

Where: EDV = Volume of grain in the endwall depth portion of the sloped section next to the endwall, bu
SD = Horizontal length of the sloped section of grain, ft
SWD = Grain depth at sidewall, ft
LPW = Level pile width, ft

Example:
Flat Storage with level
top and sloping sides.
(continued)

Equation 3.31: Total Volume of Grain in the Endwall Section

TEV = ESV + EDV

Where: TEV = Total volume of grain in the endwall section, bu
EDV = Volume of grain in the endwall depth portion of the sloped sections next to the endwall, bu
ESV = Volume of grain in the sloped sections next to the endwall, bu

Equation 3.32: Total Storage Volume

SV = CSV + TSV + TEV

Where: SV   = Storage volume of grain, bu
TEV = Total volume of grain in the endwall section, bu
TSV = Total volume of grain in the side section, bu
CSV = Volume of grain in the center section, bu

- Using Equations 3.25 to 3.32:
CSV = 28.6' x 48.6' x 16' x 0.8 bu/cu ft
= 17,791 bu

SSV = ((25.7' x 25.7' x 48.6' + (1.33 x 25.7' x 25.7' x 25.7')) x 0.47 x 0.8 bu/cu ft
= 20,558 bu

SDV = 2 x 100' x 25.7' x 4' x 0.8 bu/cu ft
= 16,448 bu

TSV = 20,558 bu + 16,448 bu
= 37,006 bu

ESV = 0.47' x 25.7' x 25.7' x 28.6' x 0.8 bu/cu ft
= 7,103 bu

EDV = 2 x 25.7' x 28.6' x 4' x 0.8 bu/cu ft
= 4,704 bu

TEV = 7,103 bu + 4,704 bu
= 11,807 bu

SV   = 17,791 bu + 37,006 bu + 11,760 bu
= 66,604 bu

- Estimate the amount of grain served by each duct location:

Equation 3.33: Volume of Grain Aerated by Center Duct

$$VGCD = \frac{CSV + TEV}{ND_{in}}$$

Where: VGCD = Volume of grain aerated by the center duct, bu
CSV   = Volume of grain in the center section, bu
TEV   = Total volume of grain in the endwall section, bu
ND_{in} = Number of ducts for inside section

VGCD = (17,791 bu + 11,807bu) / 2 ducts
= 14,799 bu/duct

Example:
Flat Storage with level
top and sloping sides.
(continued)

Equation 3.34: Volume of Grain Aerated by Outside Duct

$$VGOD = \frac{TSV}{ND_{out}}$$

Where: VGOD = Volume of grain aerated by the outside duct, bu
        TSV   = Total volume of grain in the side section, bu
        ND_{out} = Number of ducts for outside section

VGOD = 37,006 / 2 duct
       = 18,503 bu/duct

Step 3: *Calculating Airflow*
• Calculate the airflow for each duct using Equation 3.6:

  Center:
     Airflow/duct = 0.10 cfm/bu x (14,799 bu/duct)
                 = 1,480 cfm/duct
  Slope sides:
     Airflow/duct = 0.10 cfm/bu x (18,503 bu/duct)
                 = 1,850 cfm/duct

• Estimate the depth of grain over each duct and the static pressure based on the desired airflow. The depth of grain over each duct under the level center section is 16'. The depth of grain over each duct in the sloped section is found by evaluating a flat storage with peaked grain for a building width of:

Equation 3.35: Building Width for Grain Depth Calculation

W2 = 2(SOD)

Where: W2   = Building width to use for Table 3.5
       SOD = Distance from outside duct to sidewall, ft

W2 = 2(SOD) = 2 x 18'
     = 36'

• Using this building width, 36', Table 3.5 provides the depth of grain over each duct in the sloped section, 8'. Use the center duct value from Table 3.5. This value must be added to the sidewall depth. Estimate static pressure using Table 2.1:

  Center ducts:
     Grain depth   = 16.0'
     Static pressure = 0.7"

  Slope ducts:
     Grain depth   = 4.0' + 8.0'
                   = 12'
     Static pressure = 0.6"

Chapter 3

Design
Recommendations
and Procedures

Example:
Flat Storage with level
top and sloping sides.
(continued)

Step 4: *Choosing a Fan*
- Choose a fan by comparing manufacturers' data like Table 2.2.
  - Slope: Axial 12" dia.; 0.75 hp; 1,700 cfm at 1" static pressure
  - Center: Axial 12" dia.; 0.5 hp; 1,500 cfm at 1" static pressure

Step 5: *Choosing Ducts*
- Select the duct sizes from the information presented in Chapter 2, **System Components**. If the perforated duct will be round or semicircular, select the duct size using Tables 2.5 and 2.6. If the aeration fan selected moves more air than the airflow used in the calculations, select ducts that will accommodate the actual airflow provided by the fan. Maximum allowable duct air velocity is 1,500 fpm. For this example round ducts will be sized as follows using Equation 3.7:

Center duct area = 1,500 cfm/duct / 1,500 cfm/sq ft
= 1.0 sq ft
= 14" round duct

Slope duct area = 1,700 cfm/duct / 1,500 fpm
= 1.13 sq ft
= 14" round duct

Step 6: *Calculating Duct Length*
Determine the minimum perforated duct length by checking the duct surface air velocity for the perforated section of the ducts. The duct surface area is found using Table 2.5. Maximum allowable perforated duct surface air velocity is 30 fpm. Using Equation 3.8:

14" duct center:
  - Surface area = 2.93 sq ft/ft, from Table 2.5
  - Length = 1,500 cfm / (2.93 sq ft/ft x 30 fpm)
    = 17.1'

14" duct outside:
  - Surface area = 2.93 sq ft/ft
  - Length = 1,700 cfm / (2.93 sq ft/ft x 30 fpm)
    = 19.3'

- Note that the length for perforated sections as calculated in Step 1 is sufficient to meet this design criteria.

Step 7: *Sizing Exhaust Area*
- Determine the total airflow and minimum air exhaust area needed for the grain storage building using Equations 3.9 and 2.9:

Total required airflow = 1,500 cfm/duct x 2 ducts + 1,700 cfm/duct x 2 ducts
= 6,400 cfm
Exhaust area = 6,400 cfm/1,000 fpm
= 6.4 sq ft

Example:
Flat Storage with level
top and sloping sides.
(continued)

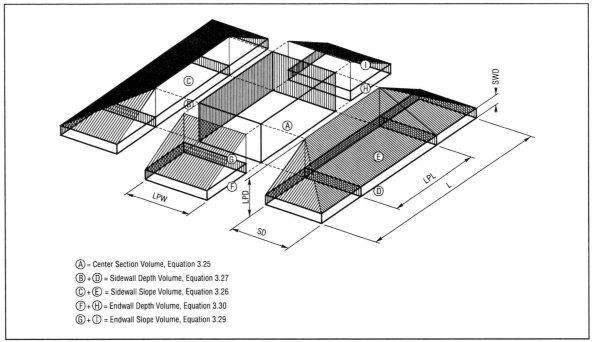

Ⓐ = Center Section Volume, Equation 3.25
Ⓑ + Ⓓ = Sidewall Depth Volume, Equation 3.27
Ⓒ + Ⓔ = Sidewall Slope Volume, Equation 3.26
Ⓕ + Ⓗ = Endwall Depth Volume, Equation 3.30
Ⓖ + Ⓘ = Endwall Slope Volume, Equation 3.29

Figure 3.12. Grain volumes of flat storage with level top and sloping sides.

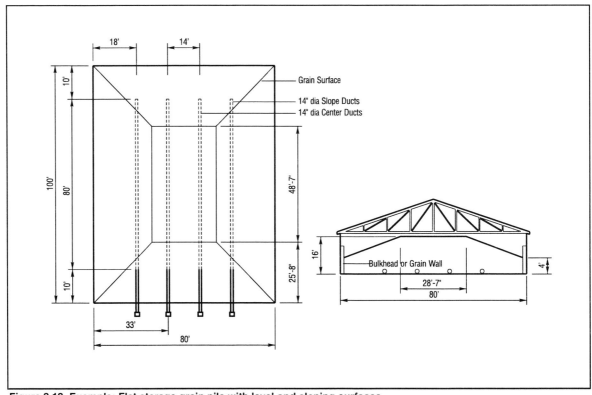

Figure 3.13. Example: Flat storage grain pile with level and sloping surfaces.

Chapter 4

Additional Examples

Example 4.1: Cylindrical Bin

SOLUTION:

Given information:
Storage dimensions: Bin Dia __36__ ft
Grain type: __Corn__ , **Grain depth:** __50__ ft
Design airflow rate: __0.10__ cfm/bu

Step 1: *Caculating Storage Volume*
- Storage Volume, Table A.4: 40,715 bu
- Total Required Airflow, Equation 1.1
    = 40,715 bu x 0.10 cfm/bu
    = 4,072 cfm

- Static Pressure, Table 2.1
    = 2.1"

Step 2: *Choosing a Fan*
- Choose a fan by comparing fan manufacturers' data like Table 2.2.
- Axial, 18" dia., 3 hp; 4,600 cfm at a static pressure of 2"

Step 3: *Choosing the System Type*
- The duct system selected is a below-floor type, Figure 2.16.

Step 4: *Determining Duct Location*
- To maintain an air distribution ratio of 1.5:1, place ducts no more than 50' apart; distance from the duct to the wall should not exceed 1/2 the grain depth, 25'. Because the bin diameter is only 36', adequate air distribution is possible with a single duct in the center.

Step 5: *Calculating Perforated Area*
- Required Perforated Area, Equation 2.7
    = 4,600 cfm / 30 fpm
    = 153 sq ft

- Duct Layout: A single aeration duct may provide the required perforated area and meet the spacing criteria outlined in Step 4.

Step 6: *Sizing the Duct*
- Duct Size: Try a 3' wide, single aeration duct, area equation from Figure A.1.
- Total perforated duct length
    = 153 sq ft / 3 ft
    = 51 ft

- This system will not work because the required duct length is greater than the bin diameter. Return to Step 5 and choose an alternative duct layout. A wider duct could be used as could a center pad design or alternative duct layouts as shown in Figure 2.11.

(Back to Step 5)
Step 5: *Determining Duct Layout*
- Duct Layout: a Y-duct or U-duct system will be used.

Step 6: *Sizing the Duct*

- Duct Size: For the Y- and U-duct systems each leg would be half the total length, 51' / 2 = 25.5'. Use 26'.

- The location of the perforated duct in the bin is determined by centering the perforated duct with equal distance between the duct ends and bin wall.

- Perforated duct cross-sectional area is determined, Equation 2.5. Each leg will carry 1/2 of the total required airflow, 2,300 cfm.

  A   = 2,300 cfm / 2,000 fpm
      = 1.15 sq ft

- Using the duct width of 3' and the square/rectangle area equation from Figure A.3:

  Perforated duct depth   = 1.15 sq ft / 3'
                          = 0.38'
                          = 5"

  Perforated Duct Dimensions: 26' x 3' x 5".

- Non-perforated supply duct cross-sectional area is determined, Equation 2.5. For the U-duct system, use the same size duct as the perforated duct to join the two legs to the single duct from the fan.

- Size the single non-perforated supply duct from the fan for both the Y- and U-duct systems.

  A   = 4,600 cfm / 2,500 fpm
      = 1.84 sq ft

- Using a 3' width, the depth is

  Non-perforated Duct Depth   = 1.84 sq ft / 3'
                              = 0.6'
                              = 8"

- Non-perforated Duct Cross-sectional Dimensions: 3' x 8".
- The perforated and non-perforated ducts normally will be the same depth; i.e. 8".

Step 7: *Sizing the Exhaust Area*

- Size Exhaust Area, Equation 2.9:

  Exhaust area = 4,600 cfm / 1,000 fpm
                = 4.6 sq ft

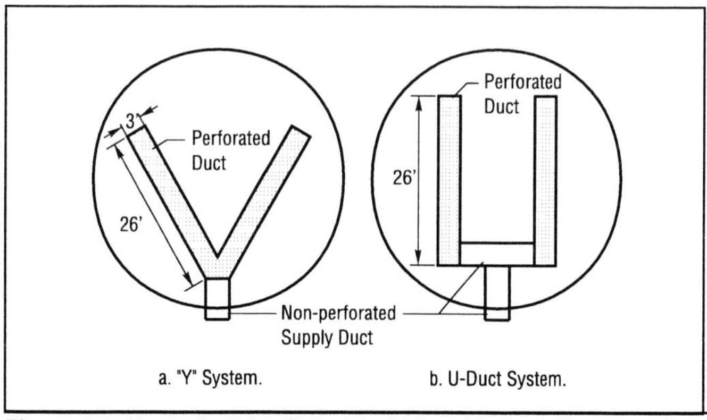

a. "Y" System.   b. U-Duct System.

Figure 4.1. Example 4.1: Cylindrical bin with flush-floor aeration.

Example 4.2: Flat Storage Level Fill

SOLUTION:

Given information:
Storage dimensions: _32_ ft x _100_ ft
Grain type: _Wheat_ , **Grain depth:** _12_ ft
Design airflow rate: _0.10_ cfm/bu

Step 1: *Calculating the Number of Ducts*
• Number of Duct Locations, Equation 2.2
>   = 32' / 12'
>   = 2.7
>   = 3 duct locations

• Duct Spacing, S1, Equation 2.3
>   = 32' / 3 duct locations
>   = 11'

• Distance of Outside Duct from Sidewall, S2, Equation 3.1
>   = 11' / 2
>   = 5'

• Center Duct Located in center of storage.

• Distance Between Duct and Endwall, E, Equation 3.2
>   = 0.7 x 5'
>   = 3.5'
>   = 4'

• Perforated Duct Length, PDL, Equation 3.3
>   = 100' - (2 x 4')
>   = 92'

Step 2: *Caculating Storage Volume*
• Storage Volume, Table 3.2: Table 3.2 does not contain values for a grain storage width of 32'. Therefore, use the equation given in Table 3.2.
>   = 32' x 100' x 12' x 0.8 bu/cu ft
>   = 30,720 bu

• Volume per Duct, Equation 3.5
>   = 30,720 bu / 3 duct
>   = 10,240 bu/duct

Step 3: *Calculating Airflow*
• Airflow per Duct, Equation 3.6
>   = 0.10 cfm/bu x 10,240 bu/duct
>   = 1,024 cfm/duct

• Static Pressure, Table 2.1
>   Static pressure = 1.0"

Step 4: *Choosing a Fan*
• Choose a fan using manufacturers' data like Table 2.2. For this example a 0.5 hp 12" axial fan moves about 1,500 cfm at a static pressure of 1", which is more than is required. Choose a smaller fan and size the ducts for this airflow. Another option would be to manifold the three ducts together and select a fan which would move 3,072 cfm at a static pressure of 1". A 1.5 hp 16" axial fan will provide the required airflow for the three ducts. For this example, the 1,500 cfm fan will be used.

Step 5: *Sizing Ducts*
• Duct Size, Equation 3.7 and Table 2.5:

Duct area = (1,500 cfm/duct) / (1,500 cfm/sq ft)
= 1.0 sq ft/duct

• Use a 14" round duct.

Step 6: *Calculating Duct Length*
• Minimum Perforated Duct Length, Table 2.5 and Equation 3.8:

Surface Area, 14" duct = 2.93 sq ft/ft
Minimum length = (1,500 cfm/duct) / (2.93 sq ft/ft x 30 fpm)
= 17.1'

• The length calculated in Step 1, 92', is sufficient to meet this design criteria.

Step 7: *Sizing Exhaust Area*
• Total Airflow, Equation 3.9

= 1,500 cfm/duct x 3 ducts
= 4,500 cfm

• Size Exhaust Area, Equation 2.9:

Exhaust area = 4,500 cfm / 1,000 fpm
= 4.5 sq ft

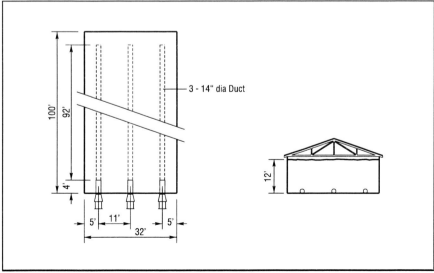

Figure 4.2. Example 4.2: 32'x100' flat storage building.
12' grain depth, level surface.

Example 4.3: Flat Storage with Peaked Grain

Given information:
Storage dimensions: _60_ ft x _100_ ft
Grain type: _Corn_ , **Sidewall grain depth:** _12_ ft,
Design airflow rate: _0.10_ cfm/bu

Step 1: *Calculating the Number of Ducts*
- Number of Duct Locations: Although this building may be served with either a one or three duct system, use a three duct system because one duct is marginal for proper aeration based on building width.
- Duct Spacing, S3, Equation 3.11
        = 60' / 4
        = 15'

- Distance of Outside Duct from Sidewall: Determined by duct spacing, S3. Center duct located in center of storage.

- Distance Between Duct and Endwall, E3, Equation 3.12
        = 0.7 x 15'
        = 10.5'
        = 11'

- Perforated Duct Length, PDL, Equation 3.3
        = 100' - (2 x 11')
        = 78'

Step 2: *Calculating Storage Volume*
- Storage Volume, Table 3.2 and 3.3, Equation 3.13:
        Total Volume   = (4,800 bu/ft x 12') + 27,072 bu
                        = 84,672 bu

- Volume per Duct, Table 3.4, Equation 3.14 and 3.15:
    Outside:
    Volume / duct   = 0.18 x 84,672 bu
                        = 15,241 bu/duct
    Center:
    Volume / duct   = 84,672 bu - (2 ducts x 15,241 bu/duct)
                        = 54,190 bu/duct

Step 3: *Calculating Airflow*
- Airflow per Duct, Equation 3.6:
    Center:
    Airflow / duct   = 0.10 cfm/bu x 54,190 bu/duct
                            = 5,419 cfm/duct
    Outside:
    Airflow / duct   = 0.10 cfm/bu x 15,241 bu/duct
                        = 1,524 cfm/duct

- Grain Depth, Table 3.5 and Equation 3.16:
    Center:
    Grain Depth   = 14' + 12'
                        = 26'

Outside:
Grain Depth   = 7' + 12'
                  = 19'
• Static Pressure, Table 2.1:
Center:
Static Pressure   = 0.9"

Outside:
Static Pressure   = 0.7"

Step 4: *Choosing a Fan*
• Choose a fan using manufacturers' data like Table 2.2.

• **Center**: Axial; 18" dia.; 3 hp; 5,700 cfm at a static pressure of 1".

• **Outside**: Axial; 12" dia.; 0.5 hp; 1,500 cfm at a static pressure of 1".

Step 5: *Sizing Ducts*
• Duct Size, Equation 3.7 and Table 2.5:
Center:
Duct area = (5,700 cfm/duct)(1,500 fpm)
          = 3.8 sq ft/duct
Use a 30" round duct.

Outside:
Duct area = (1,500 cfm/duct)(1,500 fpm)
          = 1.0 sq ft/duct

Use a 14" round duct.

Step 6: *Calculating Duct Length*
• Minimum Perforated Duct Length, Table 2.5 and Equation 3.8:
Center:
Surface area, 30" duct   = 6.28 sq ft/ft
Minimum length   = (5,700 cfm/duct)(6.28 sq ft/ft x 30 fpm)
                           = 30.3'
Outside:
Surface area, 14" duct   = 2.93 sq ft/ft
Minimum length   = (1,500 cfm/duct)(2.93 sq ft/ft x 30 fpm)
                           = 17.1'

• The length calculated in Step 1, 78' is sufficient to meet these design criteria.

Step 7: *Sizing Exhaust Area*
• Total Airflow, Equation 3.9
                           = (1,500 cfm/duct x 2 ducts) + 5,700 cfm
                           = 8,700 cfm

• Size Exhaust Area, Equation 2.9:
Exhaust Area   = 8,700 cfm / 1,000 fpm
                           = 8.7 sq ft

Flat Storage with
peaked grain
(continued)

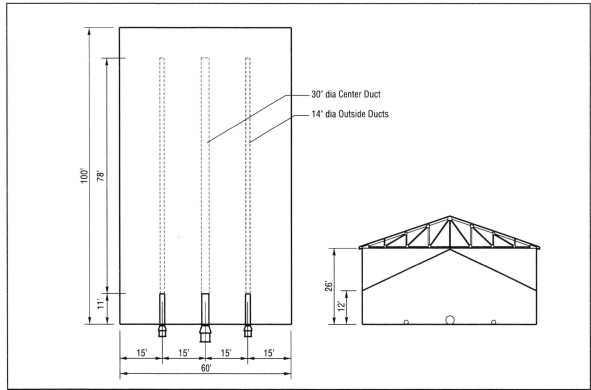

Figure 4.3. Example 4.3: 60'x100' flat storage with peaked grain.

Example 4.4: Flat Storage with Level Top and Sloping Sides

Given information:
Storage dimensions: _60_ ft x _100_ ft
Grain type: _Corn_ , **Maximum grain depth:** _14_ ft
Sidewall grain depth: _8_ ft
Design airflow rate: _0.10_ cfm/bu

Step 1: *Calculating the Number of Ducts*
• Storage Dimensions, SD, LPW, LPL, Figure 3.11:
  SD = 2.14 x (14' - 8')   Equation 3.20
       = 12.8'

  LPW = 60' - (2 x 12.8')   Equation 3.21
       = 34.4'

  LPL = 100' - (2 x 12.8')   Equation 3.22
       = 74.4'

• Number of Ducts in Level Top Portion, Equation 2.2
       = 34.4' / 14'
       = 2.5
       = 3 ducts in level surface grain

• Duct Spacing, S7, Equation 2.3
       = 34.4' / 3
       = 12'

• Distance of Second Duct from Sidewall, S6, Equation 3.24
       = 12.8' + (12' / 2)
       = 18.8'
  Use 18' to center ducts in level pile.

Distance of Outside Duct from Sidewall, S5, Equation 3.23
       = 0.7 x 12.8'
       = 9'

Distance Between end of Perforated Duct and Endwall, E5, Equation 3.12
       = 0.7 x 12
       = 9'

Perforated Duct Length, PDL, Equation 3.3
       = 100' - (2 x 9')
       = 82'

Step 2: *Calculating Storage Volume*
• Storage Volume, Equations 3.25 to 3.32 and Figure 3.12:

• CSV = 34.4' x 74.4' x 14' x 0.8 bu/cu ft
       = 28,665 bu

• SSV = ((12.8' x 12.8' x 74.4') + (1.33 x 12.8' x 12.8' x 12.8')) x 0.47 x 0.8 bu/ cu ft
       = 5,632 bu

Flat Storage with level
top and sloping sides
(continued)

- SDV = 2 x 100' x 12.8' x 8' x 0.8 bu/cu ft
  = 16,384 bu

- TSV = 5,632 bu + 16,384 bu
  = 22,016 bu

- ESV = 0.47 x 12.8' x 12.8' x 34.4' x 0.8 bu/cu ft
  = 2,119 bu

- EDV = 2 x 12.8' x 34.4' x 8' x 0.8 bu/cu ft
  = 5,636 bu

- TEV = 2,119 bu + 5,636 bu
  = 7,755 bu

- SV = 28,665 bu + 22,016 bu + 7,755 bu
  = 58,436 bu

- Volume per Duct, Equations 3.33 and 3.34:
  Center:
  Volume/duct = (28,665 bu + 7,755 bu) / 3 duct
  = 12,140 bu/duct
  Outside:
  Volume/duct = 22,016 bu / 2 duct
  = 11,008 bu/duct

Step 3: *Calculating Airflow*
- Airflow per Duct, Equation 3.6:
  Center:
  Airflow/duct   = 0.10 cfm/bu x 12,140 bu
  = 1,214 cfm/duct
  Outside:
  Airflow/duct   = 0.10 cfm/bu x 11,008 bu
  = 1,101 cfm/duct

- Grain Depth, Table 3.5, Equation 3.16 and 3.35:
  Center:
  Grain Depth   = 14'
  Outside:
  Building Width = 2 x 9'
  = 18'

- Using this building width, 18', Table 3.5 and Equation 3.16:
  Grain Depth   = 5' + 8'
  = 13'

- Static Pressure, Table 2.1:
  Center:
  Static Pressure = 0.6"

  Outside:
  Static Pressure = 0.6"

Flat Storage with level
top and sloping sides
(continued)

Step 4: *Choosing a Fan*
• Choose a fan using fan manufacturers' data like Table 2.2.
    Total fan airflow = (3 x 1,214 cfm) + (2 x 1,101 cfm) = 5,844 cfm.
    Axial, 18" dia., 3 hp; 5,700 cfm at a static pressure of 1".

Step 5: *Sizing Ducts*   Since airflow is similar in all ducts, one size will be used for all ducts.
• Duct Size, Equation 3.7 and Table 2.5:
    Airflow per duct   = 5,700 cfm / 5 = 1,140 cfm
    Duct area   = 1,140 cfm/duct / 1,500 fpm
                          = 0.8 sq ft/duct

• Use 12" round ducts.

Step 6: *Caculating Duct Length*
• Minimum Perforated Duct Length, Table 2.5 and Equation 3.8:
    Surface area, 12" duct   = 2.51 sq ft/ft
    Minimum length   = 1,140 cfm/duct / (2.51 sq ft/ft x 30 fpm)
                             = 15.1'

• The length calculated in Step 1 is sufficient to meet these design criteria.

Step 7: *Sizing Exhaust Area*
• Total Airflow, Equation 3.9
                          = 5,700 cfm

• Size Exhaust Area, Equation 2.9:
    Exhaust area   = 5,700 cfm / 1,000 fpm
                          = 5.7 sq ft

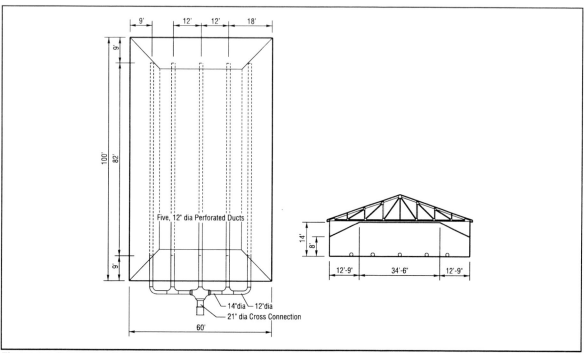

Figure 4.4. Example 4.4: 60'x100' flat storage with sloping sides and level center.

Example 4.5: Manifold

Using the results from Example 4.4 for flat storage with sloping sides and level center:

     Five 12-inch ducts with 1,140 cfm

Select non-perforated supply ducts for a manifold air distribution system that will utilize one aeration fan for the five perforated ducts.

Determine the minimum size for the manifold air ducts with a maximum air velocity of 2,500 fpm.

Solution

The duct from the fan carries all the airflow, so the required area is

    A   = 5,700 / 2,500
         = 2.28 sq ft

Table 2.5 indicates a 21-inch duct will have the needed cross-sectional area. The cross-sectional area for the non-perforated supply ducts from the center to the side ducts is

    A   = (2 x 1,140 cfm) / 2,500 fpm
         = 0.91 sq ft

Table 2.5 indicated a 14-inch duct will have the needed cross-sectional area. The non-perforated supply duct to the sloping grain duct should be the same size as the perforated duct, 12-inch diameter.

The minimum radius of curvature for the manifold is 1.5 times the pipe diameter. Therefore, the radius of curvature for the 14-inch duct is 21 inches, and the curvature for the 12-inch duct is 18 inches.

Appendices

Appendix A:
Storage Volumes, Densities, and Calculations

Table A.1. Maximum moisture contents for long-term grain storage in the Midwest.

Values for good quality, clean grain and aerated storage. Reduce 1% for poor quality grain, such as grain damaged by blight, drought, etc. For humid areas, e.g. southeastern U.S., lower moisture contents are required.

Grain type & storage time	Maximum moisture content for safe storage, %
Shelled corn and sorghum	
• Sold as #2 grain by spring	15.5
• Stored 6 to 12 months	14
• Stored more than 1 year	13
Soybeans	
• Sold by spring	14
• Stored up to 1 year	12
• Stored more than 1 year	11
Wheat, oats, barley	
• Stored up to 6 months	14
• Stored more than 6 months	13
Sunflower (oil)	
• Stored up to 6 months	10
• Stored more than 6 months	8
Flaxseed	
• Stored up to 6 months	9
• Stored more than 6 months	7
Edible beans	
• Stored up to 6 months	16
• Stored more than 6 months	4

Table A.2. Filling angles of repose and corresponding slope factors.

Filling angles of repose are affected by grain moisture content, foreign material and loading method, and can vary from these values. When estimating storage capacity, use the **average** angle.

Crop	Average		Minimum		Maximum	
	angle,	slope factor	angle,	slope factor	angle,	slope factor
Barley	28°	0.53	24°	0.45	34°	0.67
Corn	23°	0.42	21°	0.38	26°	0.49
Oats	28°	0.53	24°	0.45	32°	0.62
Grain sorghum	29°	0.55	27°	0.51	33°	0.65
Soybeans	25°	0.47	22°	0.40	29°	0.55
Sunflower						
Non-oil	28°	0.53	20°	0.36	40°	0.84
Oil	27°	0.51	18°	0.32	32°	0.62
Durum wheat	23°	0.42	22°	0.40	25°	0.47
Hard red						
Spring wheat	25°	0.47	19°	0.34	38°	0.78

Equations for calculating volumes of grain storages are shown in Figure A.1. The volume in a typical peaked flat storage building can be estimated by breaking it into three separate volumes, Figure A.2.

Use Equations A.1 to A.4 to calculate the separate volumes. Multiply the total volume, VT, by 0.8 bu/cu ft to estimate the storage capacity.

Equation A.1

$$V1 = (W)(L)(SWD)$$

Where:
V1	=	Volume of rectangular box portion
W	=	Building Width
L	=	Building Length
SWD	=	Grain depth at sidewall

Equation A.2

$$V2 = \frac{W^2 (L - W)(SF)}{4}$$

Where:
V2	=	Volume of triangular prism
W	=	Building Width
L	=	Building Length
SF	=	Slope factor, Table A.2

Equation A.3

$$V3 = \frac{W^3 (SF)}{12}$$

Where:
V3	=	Volume of pyramid
W	=	Building Width
SF	=	Slope factor, Table A.2

Equation A.4

$$VT = V1 + V2 + 2(V3)$$

Where:
VT	=	Total volume of a peaked storage
V1	=	Volume of a rectangular box
V2	=	Volume of a triangular prism
V3	=	Volume of a pyramid

Example A.1: Calculating volumes using rectangular box, triangular prisms, and pyramid equations.

Given information:
Storage dimensions: _40_ ft, x _100_ ft
Grain type: _Corn_ , **Grain depth at sidewall:** _6_ ft
Corn slope factor: _0.42_ Table A.2

Solution: Using Equations A.1 to A.4:

$$V1 = (40')(100')(6')$$
$$= 24{,}000 \text{ cu ft}$$

$$V2 = \frac{40^2 (100'- 40)(0.42)}{4}$$
$$= 10{,}080 \text{ cu ft}$$

$$V3 = \frac{40^3 (0.42)}{12}$$
$$= 2{,}240 \text{ cu ft}$$

$$VT = 24{,}000 \text{ cu ft} + 10{,}080 \text{ cu ft} + (2 \times 2{,}240 \text{ cu ft})$$
$$= 38{,}560 \text{ cu ft}$$

Storage capacity $= 38{,}560$ cu ft x 0.8 bu/cu ft
$$= 30{,}848 \text{ bu}$$

Peak height $= (1/2 \times 40' \times 0.42) + 6'$
$$= 14.4'$$

Storage Volumes,
Densities and
Calculations

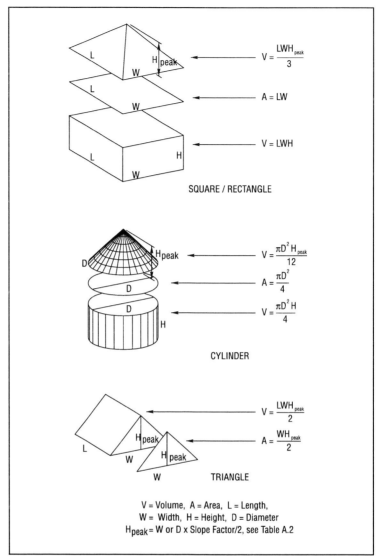

$$V = \frac{LWH_{peak}}{3}$$

$$A = LW$$

$$V = LWH$$

SQUARE / RECTANGLE

$$V = \frac{\pi D^2 H_{peak}}{12}$$

$$A = \frac{\pi D^2}{4}$$

$$V = \frac{\pi D^2 H}{4}$$

CYLINDER

$$V = \frac{LWH_{peak}}{2}$$

$$A = \frac{WH_{peak}}{2}$$

TRIANGLE

V = Volume, A = Area, L = Length,
W = Width, H = Height, D = Diameter
H_{peak} = W or D x Slope Factor/2, see Table A.2

Figure A.1. Areas and volumes.

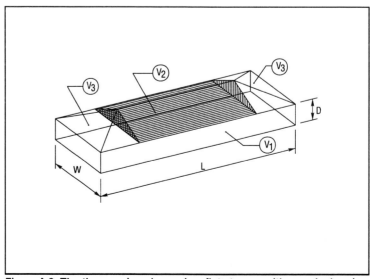

Figure A.2. The three grain volumes in a flat storage with a peaked grain surface.

Table A.3. Capacities of hopper bottom round bulk bins.

Approximate values. Overall height is the distance from ground to roof top and includes 2' of clearance between ground and hopper bottom. Roof at 35° angle. Bin volume includes roof volume. From Oklahoma State University Extension Facts No. 1103.

Bin diameter, ft	Hopper angle,°	Overall height, ft	Total volume		Volume/ft of height[a]	
			cu ft	bu	cu ft	bu
Sidedraw:						
6	60	15	227	182	28	23
		20	368	295		
		25	510	410		
9	60	22	760	611	64	51
		26	1,015	815		
		30	1,270	1,109		
Centerdraw:						
6	60	10	145	117	28	23
		15	286	230		
		20	428	344		
9	60	18	681	547	64	51
		22	935	751		
		26	1,190	955		
12	60	24	1,613	1,296	113	91
		30	2,292	1,841		
		36	2,970	2,386		
12	45	20	1,493	1,199	113	91
		30	2,624	2,108		
		40	3,755	3,016		
15	45	24	2,739	2,200	177	142
		36	4,860	3,903		
		48	6,980	5,607		
18	45	24	3,512	2,821	255	204
		36	5,268	4,231		
		48	9,619	7,726		
21	45	24	4,191	3,366	346	278
		36	8,348	6,705		
		48	12,504	10,043		

[a]For calculating volume of different height bins, add or subtract from table value.

Storage Volumes,
Densities and
Calculations

Table A.4. Estimated capacities of level full cylindrical bins.

Based on 0.8 bu = 1 cu ft. Weight of bin contents varies by actual material density, packing, and bin surface configuration.

$$bu = \frac{\pi D^2}{4} \times H \times 0.8$$

Grain depth, ft	Bin diameter, ft													
	15	18	21	24	27	30	33	36	40	42	44	46	48	60
	Bushels													
1	141	204	277	362	458	565	684	814	1,005	1,108	1,216	1,330	1,448	2,262
10	1,414	2,036	2,771	3,619	4,580	5,655	6,842	8,143	10,053	11,084	12,164	13,295	14,476	22,620
12	1,696	2,443	3,325	4,343	5,497	6,786	8,211	9,772	12,064	13,300	14,597	15,954	17,372	27,143
14	1,979	2,850	3,879	5,067	6,413	7,917	9,579	11,400	14,074	15,517	17,030	18,613	20,267	31,667
16	2,262	3,257	4,433	5,791	7,329	9,048	10,948	13,029	16,085	17,734	19,463	21,272	23,162	36,191
18	2,545	3,664	4,988	6,514	8,245	10,179	12,316	14,657	18,096	19,950	21,896	23,931	26,058	40,715
20	2,827	4,072	5,542	7,238	9,161	11,310	13,685	16,286	20,106	22,167	24,329	26,591	28,953	45,239
22	3,110	4,479	6,096	7,962	10,077	12,441	15,053	17,915	22,117	24,384	26,761	29,250	31,848	49,763
24	3,393	4,886	6,650	8,686	10,993	13,572	16,422	19,543	24,127	26,601	29,194	31,909	34,744	54,287
26	3,676	5,293	7,204	9,410	11,909	14,703	17,790	21,172	26,138	28,817	31,627	34,568	37,639	58,811
28	3,958	5,700	7,758	10,134	12,825	15,834	19,159	22,800	28,149	31,034	34,060	37,227	40,534	63,335
30	4,241	6,107	8,313	10,857	13,741	16,965	20,527	24,429	30,159	33,251	36,493	39,886	43,429	67,859
32	4,524	6,514	8,867	11,581	14,657	18,096	21,896	26,058	32,170	35,467	38,926	42,545	46,325	72,382
34	4,807	6,922	9,421	12,305	15,574	19,227	23,264	27,686	34,181	37,684	41,359	45,204	49,220	76,906
36	5,089	7,329	9,975	13,029	16,490	20,358	24,633	29,315	36,191	39,901	43,791	47,863	52,115	81,430
38	5,372	7,736	10,529	13,753	17,406	21,489	26,001	30,944	38,202	42,118	46,224	50,522	55,011	85,954
40	5,655	8,143	11,084	14,476	18,322	22,620	27,370	32,572	40,212	44,334	48,657	53,181	57,906	90,478
42	5,938	8,550	11,638	15,200	19,238	23,750	28,738	34,201	42,223	46,551	51,090	55,840	60,801	95,002
44	6,220	8,957	12,192	15,924	20,154	24,881	30,107	35,829	44,234	48,768	53,523	58,499	63,697	99,526
46	6,503	9,364	12,746	16,648	21,070	26,012	31,475	37,458	46,244	50,984	55,956	61,158	66,592	104,050
48	6,786	9,772	13,300	17,372	21,986	27,143	32,844	39,087	48,255	53,201	58,389	63,817	69,487	108,574
50	7,069	10,179	13,854	18,096	22,902	28,274	34,212	40,715	50,266	55,418	60,821	66,476	72,382	113,098
52	7,351	10,586	14,409	18,819	23,818	29,405	35,581	42,344	52,276	57,635	63,254	69,135	75,278	117,622
54	7,634	10,993	14,963	19,543	24,734	30,536	36,949	43,972	54,287	59,851	65,687	71,794	78,173	122,145
56	7,917	11,400	15,517	20,267	25,651	31,667	38,317	45,601	56,297	62,068	68,120	74,453	81,068	126,669
58	8,200	11,807	16,071	20,991	26,567	32,798	39,686	47,230	58,308	64,285	70,553	77,112	83,964	131,193
60	8,482	12,215	16,625	21,715	27,483	33,929	41,054	48,858	60,319	66,501	72,986	79,772	86,859	135,717
62	8,765	12,622	17,180	22,439	28,399	35,060	42,423	50,487	62,329	68,718	75,419	82,431	89,754	140,241
64	9,048	13,029	17,734	23,162	29,315	36,191	43,791	52,115	64,340	70,935	77,851	85,090	92,650	144,765
66	9,331	13,436	18,288	23,886	30,231	37,322	45,160	53,744	66,351	73,152	80,284	87,749	95,545	149,289
68	9,613	13,843	18,842	24,610	31,147	38,453	46,528	55,373	68,361	75,368	82,717	90,408	98,440	153,813
70	9,896	14,250	19,396	25,334	32,063	39,584	47,897	57,001	70,372	77,585	85,150	93,067	101,335	158,337
72	10,179	14,657	19,950	26,058	32,979	40,715	49,265	58,630	72,382	79,802	87,583	95,726	104,231	162,861
74	10,462	15,065	20,505	26,782	33,895	41,846	50,634	60,258	74,393	82,018	90,016	98,385	107,126	167,384
76	10,744	15,472	21,059	27,505	34,811	42,977	52,002	61,887	76,404	84,235	92,448	101,044	110,021	171,908
78	11,027	15,879	21,613	28,229	35,728	44,108	53,371	63,516	78,414	86,452	94,881	103,703	112,917	176,432
80	11,310	16,286	22,167	28,953	36,644	45,239	54,739	65,144	80,425	88,669	97,314	106,362	115,812	180,956
82	11,593	16,693	22,721	29,677	37,560	46,370	56,108	66,773	82,436	90,885	99,747	109,021	118,707	185,480
84	11,875	17,100	23,275	30,401	38,476	47,501	57,476	68,401	84,446	93,102	102,180	111,680	121,603	190,004
86	12,158	17,508	23,830	31,124	39,392	48,632	58,845	70,030	86,457	95,319	104,613	114,339	124,498	194,528
88	12,441	17,915	24,384	31,848	40,308	49,763	60,213	71,659	88,467	97,535	107,046	116,998	127,393	199,052
90	12,723	18,322	24,938	32,572	41,224	50,894	61,582	73,287	90,478	99,752	109,478	119,657	130,288	203,576
92	13,006	18,729	25,492	33,296	42,140	52,025	62,950	74,916	92,489	101,969	111,911	122,316	133,184	208,100
94	13,289	19,136	26,046	34,020	43,056	53,156	64,319	76,544	94,499	104,186	114,344	124,975	136,079	212,623
96	13,572	19,543	26,601	34,744	43,972	54,287	65,687	78,173	96,510	106,402	116,777	127,634	138,974	217,147
98	13,854	19,950	27,155	35,467	44,888	55,418	67,056	79,802	98,521	108,619	119,210	130,293	141,870	221,671
100	14,137	20,358	27,709	36,191	45,805	56,549	68,424	81,430	100,531	110,836	121,643	132,953	144,765	226,195

Table A.4. Grain bulk seed densities.

Based on weights and measures used by the U.S. Department of Agriculture. Adapted from ASAE Data: ANSI/ASAE D241.4 FEB 93. 1 bu = 1.244 ft^3

Grain or seed	Approximate net weight[a], lb/bu
Alfalfa	60
Barley	48
Beans:	
Lima, dry	56
Lima, unshelled	28-32
Snap	28-32
Other, dry (navy, pinto, etc.)	60
Bluegrass	14-30
Broomcorn seed	44-50
Buckwheat	48
Canola (rapeseed)	52
Castor beans	41
Clover seed	60
Corn:	
Ear, husked	70[b]
Shelled	56
Cottonseed	32
Cowpeas	60
Flaxseed	56
Hempseed	44
Kapok seed	35-40
Lentils	60
Millet	48-50
Mustard seed	58-60
Oats	32
Orchard grass seed	14
Peanuts, unshelled:	
Virginia type	17
Runners, Southeastern	21
Spanish	25
Perilla seed	37-40
Popcorn:	
On ear, husked	70[b]
Shelled	56
Poppy seed	46
Redtop seed	27-35
Rice, rough	45
Rye	56
Sesame	46
Sorgo seed	50
Sorghum grain	56
Soybeans	60
Spelt (p. wheat)	40
Sudan grass seed	40
Sunflower seed, non-oil	24
Sunflower seed, oil	32
Timothy seed	45
Velvet beans (hulled)	60
Vetch	60
Walnuts, black	50
Wheat	60

[a]Net weight is the standard bushel weight if one has been defined.

[b]70 lb of husked, ear corn yields 1 bu, or 56 lb of shelled corn. 70 lb ear corn occupies 2 volume bu (2.5 cu ft).

Table A.6. Fraction/Decimal Equivalents.

Fraction		Decimal
1/30	=	0.03
1/20	=	0.05
1/10	=	0.10
1/8	=	0.125
1/4	=	0.25
1/5	=	0.20
1/2	=	0.50
3/4	=	0.75
1	=	1.0

Appendix B: References

The following sources have been used in part to produce this publication.

- ASAE Standards, 1997. American Society of Agricultural Engineers, 2950 Niles Road, St. Joseph, MI.
  D272.3. *Resistance to Airflow of Grains, Seeds, Other Agricultural Products, and Perforated Metal Sheets.*
  EP 433. *Loads Exerted by Free-Flowing Grain on Bins.*
  EP 545. *Loads Exerted by Free-Flowing Grains on Shallow Storage Structures.*

- Bloome, et. al. 1975. *Aeration System Design for Cone-Bottom Round Bins.* OSU Extension Facts 1103, Oklahoma State University, Stillwater, OK.

- Brooker, Bakker-Arkema, Hall. 1992. *Drying and Storage of Grains and Oilseeds.* Van Nostrand Reinhold, New York, NY.

- Burrell. 1974. Aeration. In: *Storage of Cereal Grains and Their Products.* Amer. Assoc. Cereal Chemists. St. Paul, MN.

- Chapman, Morey, Cloud and Nieber. 1989. *Airflow patterns in flat storage aeration systems.* Trans. ASAE 32(4)1368.

- Cloud and Morey. 1980. *Fan and equipment selection for natural-air drying, dryeration, in-storage cooling, and aeration systems.* Bulletin M-166. Univ. Minnesota, St. Paul, MN.

- Foster and Stahl. 1959. *Operating Grain Aeration Systems in the Corn Belt.* Marketing Research Report No. 337, United States Department of Agriculture, Agricultural Marketing Service, Marketing Research Division, Washington, D.C.

- Foster and Tuite. 1992. *Aeration and Stored Grain Management.* In: *Storage of Cereal Grains and Their Products.* Amer. Assoc of Cereal Chemists. St. Paul, MN.

- Hellevang. 1993. *Natural Air/Low Temperature Crop Drying.* Extension Bulletin 35 (revised), July, 1993. NDSU Extension Service, Fargo, ND.

- Holman. 1960. *Aeration of Grain in Commercial Storages.* MRR-178. USDA.

- Kline and Converse. 1961. *Operating Grain Aeration Systems in the Hard Winter Wheat Area.* Marketing Research Report No. 480, Transportation and Facilities Research Division, Agricultural Marketing Service, United States Department of Agriculture in cooperation with the Kansas Agricultural Experiment Station.

- Watson. 1987. *Aeration System Design for Flat Grain Storages Utilizing Interactive Computer Graphics.* Ph.D. thesis, Michigan State University.

Additional sources from MidWest Plan Service, 122 Davidson Hall, Iowa State University, Ames, IA 50011-3080.
  Managing Dry Grain in Storage, AED-20. 1980.
  Grain Drying, Handling and Storage Handbook, MWPS-13, 2nd edition. 1987.
  Low Temperature and Solar Grain Drying Handbook, MWPS-22, 1st edition. 1983.

Appendix C: Glossary

bu:

Bushel. For the purposes of design, a bushel refers to a **volume** of 1.244 cubic feet regardless of the type of grain. Other "bushels" include **dry** bushel: the weight of grain at a standard moisture content as determined by USDA. **Weight** bushel is the weight of grain defined by USDA to be a bushel regardless of moisture content.

cfm:

Cubic feet per minute. A measure of airflow volume supplied by a fan per period of time.

cfm/bu:

Cubic feet of air per minute per bushel. A measure of airflow supplied per volume of aerated grain. Divide total cubic feet of air supplied per minute by total bushels of grain in the storage structure.

cfm/sq ft:

Cubic feet per minute per square foot. A measure of airflow supplied per square foot of storage floor area, duct perforated area, or duct cross-sectional area. Divide fan airflow supplied by the square footage of storage floor area, perforated area, or duct cross-sectional area.

dry grain:

Grain at the recommended long-term storage moisture content, Table A.1.

fpm:

Feet per minute. A measure of air velocity. A velocity of 88 fpm = 1 mile per hour (mph).

friction loss:

The resistance of sliding over a surface. In an aeration duct, it is the effect of the roughness of the duct that impedes the movement of the air.

moisture migration:

The movement of moisture within the stored grain due to convection air currents caused by temperature variations within the stored grain. Air near the bin wall is cooled, sinks to the bottom of the bin, and pushes up warm air in the center of the bin. Moisture from the air is deposited on the grain as the air is cooled near the top center of the stored grain.

static pressure:

Measured in inches of water, is the force per unit area a fan must develop to overcome the resistance to airflow created by grain and equipment.

specific heat:

the energy required to change the temperature of one pound of mass by one degree.

test weight:

is a quality factor (lb/bu) referring to the weight of the grain that fits into a volume bushel. This is measured indirectly by a prescribed USDA method using a filling hopper and a quart cup (1/32 of bu). See Table A.4.

Appendix D: List of Tables, Design Guide, Design Equations

List of Tables

Design Guide for Flat Storages

Given Information:
Storage dimensions: _____
Grain type: _____, **Sidewall Grain depth:** ___ ft
Maximum Grain depth: ___ ft
Design airflow rate: _____ cfm/bu

			Peaked Filled		
	Level Fill	0 to 60 feet wide	60 to 100 feet wide	Crosswise Duct	Level Fill, sloping sides
Step 1					
Number of duct locations	Equation 2.2, pg. 13	1	3	Equation 3.18, pg. 49	Equation 2.2, pg. 13
Duct Spacing, ft	Equation 2.3, pg. 13	centerline	Equation 3.11, pg. 40	Equation 3.17, pg. 48	Equation 2.3, pg. 13
					Equation 3.24, pg. 52
Distance of outside duct from sidewall, ft	Equation 3.1, pg. 36	———— determined by duct spacing ————			Equation 3.23, pg. 52
Distance between duct and endwall/sidewall, ft	Equation 3.2, pg. 36	Equation 3.10, pg. 40	Equation 3.12, pg. 40	Equation 3.12, pg. 40	Equation 3.12, pg. 40
Perforated duct length, ft	Equation 3.3, pg. 37	Equation 3.3, pg. 37	Equation 3.3, pg. 37	Equation 3.5, pg. 37	Equation 3.3, pg. 37
Step 2					
Storage Volume, bu	Equation 3.4, pg. 37	Equation 3.13, pg. 44	Equation 3.13, pg. 44	Equation 3.13, pg. 44	Equation 3.25 to Equation 3.32, pg. 53-55
Volume of grain aerated per duct, bu	Equation 3.5, pg. 37	Equation 3.13, pg. 44	Equation 3.14, pg. 44	Equation 3.5, pg. 37	Equation 3.33, pg. 55
			Equation 3.15, pg. 44		Equation 3.34, pg. 56
Step 3					
Airflow per duct, cfm	Equation 3.6, pg. 37	Equation 3.6, pg. 37	Equation 3.6, pg. 37	Equation 3.6, pg. 37	Equation 3.6, pg. 37
Grain Depth, ft	given	Equation 3.16, pg. 45	Equation 3.16, pg. 45	Equation 3.19, pg. 50	Equation 3.35, pg. 56
					Equation 3.16, pg. 45

Static Pressure — Table 2.1, pg. 7; Use fan airflow for the following steps.

Step 4
Choosing a Fan from Manufacturer's Data (Similar to Table 2.2)

Step 5

Duct size, sq ft or diameter — Equation 3.7, pg. 38

Step 6
Minimum perforated duct length, ft — Equation 3.8, pg. 38

Step 7
Total required airflow, cfm — Equation 3.9, pg. 39
Size exhaust are, sq ft — Equation 2.9, pg. 21

Design Equations

Equation	Number	Page

$Q = (GV)(AR)$ **Required airflow**
 	
Where: Q   =   Total airflow, cfm
          GV =   Volume of grain, bu
          AR =   Airflow rate, cfm/bu
	1.1	2

$AT = \dfrac{15}{AR}$ (for grain density of 60 lb/bu) **Aeration time (60lb/bu)** Where: AT = Aeration time, hrs AR = Airflow rate, cfm/bu	1.2	3
$AT = \left(\dfrac{TW}{60}\right)\dfrac{15}{AR}$ **Aeration time (test weight)** Where: AT = Aeration time, hrs TW = Test weight, lb/bu	1.3	3

$EFMHP = \dfrac{(Q)(SP)}{6,350(SE)}$
             (for grain density of 60 lb/bu)   **Estimated fan horsepower**
 	
Where: EFMHP =   Estimated fan motor horsepower, hp
          Q   =   Total airflow, cfm
          SP   =   Static pressure, inches of water
          SE   =   Static efficiency, decimal. Ratio of the theoretical power
                      needed to move the air against the static pressure to the
                      actual power of the motor. Typical values range from 0.40 for
                      lower   horsepower fans to 0.70 for higher horsepower fans.
	2.1	9

$ND = \dfrac{W \text{ or } BD}{GD}$ **Number of ducts, level fill**
Where: ND =   Number of ducts
          BD =   Bin diameter, ft
          W   =   Building width, ft
          GD =   Grain depth, ft
	2.2	13

$SL = \dfrac{W \text{ or } BD}{ND}$ **Duct spacing, level fill**
Where: SL =   Duct spacing, level fill, ft
          ND =   Number of ducts
          BD =   Bin diameter, ft
          W   =   Building width, ft
	2.3	13

$Q = (A)(V)$ **Calculated airflow**
Where: Q =   Total airflow, cfm
          A =   Duct cross-sectional area, sq ft
          V =   Air velocity, fpm
	2.4	15

$A = \dfrac{Q}{V}$ **Duct area**
Where: Q =   Total airflow, cfm
          A =   Duct cross-sectional area, sq ft
          V =   Air velocity, fpm
	2.5	15

$V = \dfrac{Q}{A}$ **Air velocity**
Where: Q =   Total airflow, cfm
          A =   Duct cross-sectional area, sq ft
          V =   Air velocity, fpm
	2.6	15

$AS = \dfrac{Q}{VS}$ **Perforated surface area**
Where: VS =   Surface air velocity, fpm
          Q   =   Total airflow, cfm
          AS =   Perforated surface area, sq ft
          $AS = \pi(D)(DL)\,(0.8)$ (round ducts)
          $AS = \dfrac{\pi(D)(DL)}{2}$ (semicircular ducts)
          D   =   Duct diameter, ft
          DL   =   Duct length, ft
          π =   3.1416 (pi)
	2.7	15

$PL = BD - GD$ **Aeration pad length**
Where: PL =   Pad length, ft
          BD =   Bin diameter, ft
          GD =   Grain depth, ft
	2.8	16

Design Equations

Equation				Number	Page
$AV = \dfrac{Q}{1,000}$			**Vent area**	2.9	21
Where:	AV	=	Minimum vent area, sq ft		
	Q	=	Total airflow, cfm		
$EO = \dfrac{\pi(BD)(EW)}{12}$			**Bin eave opening**	2.10	21
Where:	BD	=	Bin diameter, ft		
	EO	=	Eave opening, sq ft		
	EW	=	Width of eave opening, in		
$SDW = \dfrac{SL}{2}$			**Side duct spacing**	3.1	36
Where:	SDW	=	Distance from sidewall to first duct, ft		
	SL	=	Duct spacing, ft		
$EP = 0.7(S2)$			**Endwall duct spacing, level fill**	3.2	36
Where:	EP	=	Distance from endwall to perforated duct, ft		
	SDW	=	Distance from sidewall to first duct, ft		
$PDL = L - 2(EP)$			**Perforated duct length**	3.3	37
Where:	PDL	=	Perforated duct length, ft		
	L	=	Building length, ft		
	EP	=	Distance from endwall to perforated duct, ft		
$TV = (VG)(GD)$			**Storage volume, level fill**	3.4	37
Where:	TV	=	Total storage volume, bu		
	VG	=	Volume of grain per foot of depth, bu/ft (Table 3.2)		
	GD	=	Grain depth, ft		
$GAD = \dfrac{TV}{ND}$			**Grain aerated per duct**	3.5	37
Where:	GAD	=	Amount of grain aerated per duct, bu		
	TV	=	Total storage volume, bu		
	ND	=	Number of ducts		
$APD = (AR)(GAD)$			**Airflow per duct**	3.6	37
Where:	APD	=	Airflow per duct, cfm		
	GAD	=	Amount of grain aerated per duct, bu		
	AR	=	Airflow rate, cfm/bu		
$DA = \dfrac{APD}{1,500}$			**Recommended duct area**	3.7	38
Where:	DA	=	Recommended duct area, sq ft		
	APD	=	Airflow per duct, cfm		
$PDL_{min} = \dfrac{APD}{(SA)(30)}$			**Perforated duct length**	3.8	38
Where:	PDL_{min}	=	Minimum perforated duct length, ft		
	SA	=	Surface area, sq ft/ft length		
	APD	=	Airflow per duct, cfm		
$Q_{req'd} = (APD)(ND)$			**Total airflow**	3.9	39
Where:	$Q_{req'd}$	=	Total airflow, cfm		
	APD	=	Airflow per duct, cfm		
	ND	=	Number of ducts		
$EPP = 0.7\dfrac{W}{2}$			**Endwall duct spacing, peaked grain, one duct, w < 60'**	3.10	40
Where:	EPP	=	Distance from endwall to perforated duct, ft		
	W	=	Building width, ft		
$S3 = \dfrac{W}{4}$			**Three-duct spacing, peaked grain, W < 100'**	3.11	40
Where:	S3	=	Spacing between ducts, ft		
	W	=	Building width, ft		
$E3 = 0.7(S3)$			**Endwall duct spacing for 3-duct design, peaked grain, W < 100'**	3.12	40
Where:	E3	=	Distance from endwall to perforated duct, ft		
	S3	=	Spacing between ducts, ft		
$SV = (BLV)(SWD) + (PV)$			**Peaked storage volume**	3.13	44
Where:	SV	=	Storage volume, bu		
	BLV	=	Bottom Level Volume, bu (Table 3.2)		
	SWD	=	Sidewall Depth, ft		
	PV	=	Peaked volume, bu (Table 3.3)		

Equation	Number	Page

VGOD = C1(SV)   **Peaked volume aerated by outside duct**
Where: VGOD   =   Volume of grain aerated by outside duct, bu/duct
             C1   =   Coefficient from Table 3.4
             SV   =   Storage volume, bu
	3.14	44

VGCD = SV - 2(VGOD)   **Peaked volume aerated by corner duct**
Where: VGCD   =   Volume of grain aerated by center duct, bu/duct
            VGOD =   Volume of grain aerated by outside duct, bu/duct
             SV   =   Storage volume, bu
	3.15	44

GD = GDP + SWD   **Grain depth over ducts**
Where: GD   =   Grain depth, ft
             GDP   =   Peaked grain depth over duct, ft (Table 3.5)
             SWD   =   Grain depth at sidewall, ft
	3.16	45

SC = 2 + 2(SWD) **Crosswise duct spacing** Where: S4 = Duct spacing for crosswise duct layout, ft SWD = Grain depth at sidewall, ft	3.17	48

$ND = \dfrac{L}{S4} - 1$ **Number of crosswise ducts**
Where: ND   =   Number of ducts
             S4   =   Duct spacing for crosswise duct layout, ft
             L   =   Building length, ft
	3.18	49

$GD_{avg} = \dfrac{GDP}{2} + SWD$ **Average grain depth**
Where: GD_{avg} =   Average grain depth, ft
            GDP =   Peak grain depth, ft (Table 3.5)
            SWD   =   Grain depth at sidewall, ft
	3.19	50

SD = 2.14(LPD - SWD)   **Sloped pile distance**
Where: SD   =   Horizontal length of the sloped section of grain, ft
             LPD   =   Level pile depth, ft
              SWD   =   Grain depth at sidewall, ft
	3.20	52

LPW = W - 2(SD)   **Level pile width**
Where: LPW   =   Level pile width, ft
             SD   =   Horizontal length of the sloped section of grain, ft
             W   =   Building width, ft
	3.21	52

LPL = L - 2(SD)   **Level pile length**
Where: LPL   =   Level pile length, ft
             L   =   Building length, ft
             SD   =   Horizontal length of the sloped section of grain, ft
	3.22	52

SOD = 0.7(SD) **Outside duct spacing** Where: SOD = Distance from outside duct to sidewall, ft SD = Horizontal length of the sloped section of grain, ft	3.23	52

$S2D = SD + \dfrac{SLP}{2}$ **Second duct spacing**
Where: S2D   =   Distance from second duct to sidewall, ft
            SLP   =   Duct spacing for the level section of grain, ft
            SD   =   Horizontal length of the sloped section of grain, ft
	3.24	52

CSV = 0.8(LPW)(LPL)(LPD)   **Center section volume**
Where: CSV   =   Volume of grain in the center section, bu
            LPW   =   Level pile width, ft
            LPL   =   Level pile length, ft
            LPD   =   Level pile depth, ft
	3.25	53

Sloped side section volume
$SSV = 0.8(0.47)\left[\left(SD^2\right)(LPL) + \left(\dfrac{4\left(SD^3\right)}{3}\right)\right]$
Where: SSV   =   Volume of grain in the sloped side sections, bu
            SD   =   Horizontal length of the sloped section of grain, ft
            LPL   =   Level pile length, ft
	3.26	54

SDV = (2)(0.8)(L)(SD)(SWD)   **Sidewall depth volume**
Where: SDV   =   Volume of grain in the the sidewall height portion of sloped section, bu
            SD   =   Horizontal length of the sloped section of grain, ft
            SWD   =   Grain depth at sidewall, ft
            L   =   Building length, ft
	3.27	54

Peaked Flat Storage, Crosswise Ducts

Flat Storage, Level Top, Sloping Sides

Equation	Number	Page

TSV = SSV + SDV   **Total side volume**
Where: TSV = Total volume of grain in the side section, bu
        SSV = Volume of grain in the sloped side sections, bu
        SDV = Volume of grain in the sidewall height portion
                  of the sloped section, bu
	3.28	54

$ESV = 0.8(0.47)(SD^2)(LPW)$ **Endwall slope volume**
Where: ESV = Volume of grain in the sloped sections next to endwall, bu
        SD   = Horizontal length of the sloped section of grain, ft
        LPW = Level pile width, ft
	3.29	54

EDV = 2(0.8)(SD)(LPW)(SWD)   **Endwall depth volume**
Where: EDV = Volume of grain in the endwall depth portion
                  of the sloped sections next to the endwall, bu
        SD   = Horizontal length of the sloped section of grain, ft
        SWD = Grain depth at sidewall, ft
        LPW = Level pile width, ft
	3.30	54

TEV = ESV + EDV   **Total endwall volume**
Where: TEV = Total volume of grain in the endwall section, bu
        EDV = Volume of grain in endwall depth portion of the sloped sections
                  next to the endwall, bu
        ESV = Volume of grain in the sloped sections next to endwall, bu
	3.31	55

SV = CSV + TSV + TEV   **Storage volume**
Where: SV   = Storage volume of grain, bu
        TEV = Total volume of grain in the endwall section, bu
        TSV = Total volume of grain in the side section, bu
        CSV = Volume of grain in the center section, bu
	3.32	55

$VGCD = \dfrac{CSV + TEV}{ND_{in}}$ **Grain aerated by center duct**
Where: VGCD = Volume of grain aerated by the center duct, bu
        CSV   = Volume of grain in the center section, bu
        TEV   = Total volume of grain in the endwall section, bu
        ND_{in} = Number of ducts for inside section
	3.33	55

$VGOD = \dfrac{TSV}{ND_{out}}$ **Grain aerated by outside duct**
Where: VGOD = Volume of grain aerated by the outside duct, bu
        TSV   = Total volume of grain in the sidewall section, bu
        ND_{out} = Number of ducts for outside section
	3.34	56

W2 = 2(S5) **Building width for depth** Where: W2 = Building width to use for Table 3.5, ft **calculation** S5 = Distance from outside duct to sidewall, ft	3.35	56

V1 = (W)(L)(SWD)
Where: V1 = Volume of rectangular box portion
        W   = Building Width
        L   = Building Length
        SWD = Grain depth at sidewall
	A.1	73

$V2 = \dfrac{(W^2)(L - W)(SF)}{4}$
Where: V2 = Volume of rectangular box portion
        W   = Building Width
        L   = Building Length
        SF   = Slope factor, Table A.2
	A.2	73

$V3 = \dfrac{(W^3)(SF)}{12}$
Where: V3 = Volume of a pyramid
        W   = Building Width
        SF   = Slope factor, Table A.2
	A.3	73

VT = V1 - V2 + 2(V3)
Where: VT = Total volume of a peaked storage
        V1 = Volume of a rectangular box
        V2 = Volume of triangular prism
        V3 = Volume of pyramid
	A.4	73

Flat Storage, Level Top, Sloping Sides

Volume in Peaked Flat Storage

Appendix E: Variable Definitions

A	=	Duct cross-sectional area, sq ft
APD	=	Airflow per duct, cfm
AR	=	Airflow rate, cfm/bu
AS	=	Perforated duct surface area, sq ft
AT	=	Aeration time, hrs
AV	=	Minimum vent area, sq ft
BD	=	Bin diameter, ft
BLV	=	Bottom level volume, from Table 3.2
BHP	=	Power delivered to the fan by motor shaft, hp
C1	=	Coefficient from Table 3.4
CSV	=	Volume of grain in the center section, bu
D	=	Duct diameter, ft
DA	=	Duct area, sq ft
DL	=	Duct length, ft
EP	=	Distance from endwall to perforated duct, ft
EPP	=	Distance from endwall to perforated duct, peaked grain, ft
E3	=	Distance from endwall to perforated duct, peaked grain, three ducts, ft
EDV	=	Volume of grain in the endwall depth portion of the sloped sections next to the endwall, bu
EFMHP	=	Estimated fan motor horsepower, hp
EO	=	Eave opening, sq ft
EPC	=	Distance from sidewall to perforated duct, crosswise ducts, ft
ESV	=	Volume of grain in the sloped sections next to the endwall, bu
EW	=	Width of eave opening, in
GAD	=	Amount of grain aerated per duct, bu
GD	=	Grain depth, ft
GD_{avg}	=	Average grain depth, ft
GDP	=	Peaked grain depth, ft
GV	=	Volume of grain, bu
L	=	Building length, ft
LPD	=	Level pile depth, ft
LPL	=	Level pile length, ft
LPW	=	Level pile width, ft
ND	=	Number of ducts
PDL	=	Perforated duct length, ft
PDL_{min}	=	Minimum perforated duct length, ft
PL	=	Pad length, ft
Q	=	Total airflow, cfm
$Q_{req'd}$	=	Required airflow, cfm
SL	=	Duct spacing, level fill, ft
SDW	=	Distance from sidewall to duct, ft
S3	=	Spacing between ducts, three duct system, ft
SC	=	Duct spacing for crosswise duct layout, ft
SOD	=	Distance from outside duct to sidewall, ft
S2D	=	Distance from second duct to sidewall, ft
SLP	=	Duct spacing for the level section of grain, ft
SA	=	Surface area, sq ft/ft length
SD	=	Horizontal length of the sloped section of grain, ft
SDV	=	Volume of grain in the sloped section in the area of the sidewall height, bu
SE	=	Static efficiency, decimal.
SHP	=	Static air horsepower, hp
SP	=	Static pressure, inches of water
SSV	=	Volume of grain in the sloped side sections, bu
SV	=	Storage volume, bu
SWD	=	Grain depth at sidewall, ft
TEV	=	Total volume of grain in the sloped section of the endwall, bu
TSV	=	Total volume of grain in the side section, bu
TV	=	Total storage volume, bu
TW	=	Test weight, lb/bu
V	=	Air velocity, fpm
VG	=	Volume of grain per foot of depth, bu/ft
VGCD	=	Volume of grain aerated by center duct, bu/duct
VGOD	=	Volume of grain aerated by outside duct, bu/duct
VS	=	Duct surface air velocity, fpm
W	=	Building width, ft

Index